Antionette V.H. Wakeman

Scientific Sewing and Garment Cutting

For use in schools and in the home

Antionette V.H. Wakeman

Scientific Sewing and Garment Cutting
For use in schools and in the home

ISBN/EAN: 9783337418359

Printed in Europe, USA, Canada, Australia, Japan

Cover: Foto ©berggeist007 / pixelio.de

More available books at **www.hansebooks.com**

AND

GARMENT CUTTING

For Use in Schools and in the Home

BY

ANTOINETTE VAN HOESEN WAKEMAN

AND

LOUISE M. HELLER

SILVER, BURDETT & COMPANY

NEW YORK   BOSTON   CHICAGO

PREFACE.

THIS work on *Scientific Sewing and Garment Cutting* owes its publication to the constant and increasing demand for information in regard to the system which it explains. This demand has been created by the unqualified success of this form of manual training in the school where it has been taught, substantially as here set forth, for the past six years.

Since it is not a theory reduced to possible practice, but the exposition of a system that has been productive of the most excellent results, it is given to the public in the confident belief that it will serve a useful purpose.

PUBLISHERS' NOTE.

It will be noted in connection with the diagrams presented in this book, that the authors have indicated lines by a single letter. This is for conciseness; and no confusion need arise if the general scheme of the parallelogram be borne in mind, which in every case has for its base line A, its left-hand side B, the upper line C, and the right-hand side D. In the more complex figures arrows are used to designate the direction of the lines.

The publishers are permitted to announce that worked models described in this book may be obtained by addressing Miss Louise M. Heller, 293 South Oakley Avenue, Chicago, Illinois.

CONTENTS.

5

CHAPTER III.

SECOND GRADE WORK.

CHAPTER IV.

THIRD GRADE WORK.

CHAPTER V.

FOURTH GRADE WORK.

CHAPTER VI.
WORK OF THE FIFTH GRADE.

CHAPTER VII.
SIXTH GRADE WORK.

CHAPTER VIII.
SEVENTH GRADE WORK.

CHAPTER IX.

EIGHTH GRADE WORK.

INTRODUCTION.

THE system of instruction set forth in this book makes sewing and garment cutting an educational factor identical with manual training. It has been the primary aim of the authors to lead the pupils to think independently, coördinately, and constructively. To this end the reason for each step in the course of instruction has been set forth explicitly, and the teacher is urged to make these reasons plain to the pupils, that they may work from intelligent conviction, and not mechanically. To fail in this is to defeat the first and most important aim of the system, which is founded not only upon broad educational principles, but upon mathematical verities.

The entire system is based upon the square and the parallelogram, and in this respect the sewing is coördinate with the garment cutting. The system of cutting, which is without chart or other guide than simple, easily comprehended mathematical principles, is original with Miss Louise M. Heller. For six years Miss Heller has been connected with the department of sewing and garment cutting in the Chicago Jewish Manual Training-School; and this system, which is now for the first time given to the public, has been thoroughly tested in that institution.

It is Huxley who claims that that person is liberally educated who has been so trained that his body is the ready servant of his will, and does with ease and pleasure all the work which, as a mechanism, it is capable of doing. That the system which is clearly explained in the following pages is a valuable factor in obtaining this result, has been abundantly proved in the school where it has been tested. In this institution it is not the aim of the course to graduate proficient seam-

9

stresses, but rather to so educate pupils that they may be able to make the most of themselves in any one of many lines into which opportunity and their capacity may lead them. It is a fact that perfect drafting and the most satisfactory needlework are done with the greatest ease by those pupils who have taken the course of sewing and garment cutting. It is a matter of daily, nay, of hourly, experience in this school that a girl of ten years takes the measures and drafts in five minutes the pattern of a perfectly fitting dress waist for a child. Other garments are drafted and cut with equal ease by the aid of this system.

The counting of threads, the accurate measurements required, the precision in the matter of darning, and all else that pertains to the work of the course, may seem to the superficial observer an unnecessary expenditure of time and effort; but let it be borne in mind that the first aim of the system is to enable the child to see correctly, to use what she sees with facility, and to make her hand the ready servant of her will. More than this, it is easy to demonstrate that, when the habit of executing the commonest task in the best way is established, the work can be done with no greater expenditure of time and effort than is employed in doing it in the least skillful manner.

In all lines, the artistic, which until recently has been appropriated by exclusive classes, is now being incorporated into the common every-day life of the people, and the work of the needle should be no exception. Knowledge alone is required to accomplish this; for the skillful use of common materials, which are among daily necessities, renders the work of the needle truly artistic.

The stories of materials and their uses, which follow the outline of work for each grade, are not a part of the course. They can be used or not at the discretion of the teacher, and must be adapted by her to the pupils she is instructing. All that has been attempted is to give a general idea of the different subjects in a form suited, as far as possible, to the capacity of the children of the respective grades.

It may seem that, as the utmost precision is required at every step, from the beginning to the end of the course, the creative faculty in the child is not sufficiently encouraged; but let it be remembered that when principles are thoroughly mastered, the worker is made free. The small squares of the canvas of the first models represent the regular stitch; and having mastered this, the pupil is enabled to work with ease along correct lines. It is one thing to create, and quite another thing to produce, that which is of value; and it is only when those principles which exist in the nature of things are recognized and obeyed that real excellence is achieved.

It is true that the child is required to do certain given tasks in a certain way; but she has no set pattern, and really designs each model without assistance save direction from the teacher. The working out of the different designs in this way establishes in the pupil's mind that most valuable of possessions, a correct ideal.

It will be observed that there is no sewing up and down through cards perforated in formal designs, for this kind of work is of very little practical value; it is mechanical, and in doing it the child acquires habits which must be overcome later.

It is an axiom of modern pedagogics that no portion of that fine and complex instrument, the human body, should be neglected; and, keeping this in mind, the authors have taken many things into account in putting forth their system. Throughout the course the appeal is made, not to the pupil's memory, but to her understanding. Although but forty minutes twice a week are devoted to the work, it has been found that this course invariably stimulates the reasoning faculties, and brings into action powers of the mind previously dormant.

While *Scientific Sewing and Garment Cutting* is arranged as a text-book for schools, it is also a valuable manual for the home circle. Whether used as a guide in cutting and making garments for children, or in teaching children to sew, or as a handbook containing much useful and interesting general information, it is one that every mother of a family will find of value.

CLOTHING AND ITS USES.

There is no authentic history of the beginning of sewing, neither is there any detailed account of the various stages of clothing, although it is certain that the skins of beasts take precedence of all other material as wearing apparel. Skins furnished the winter garb of the Briton, and supplied the covering of the wild tribesmen that followed the hosts of Xerxes in his expeditions against Hellas. From those remote times until the present, the skins of animals have been used in various ways by all classes and conditions of men for garments.

The garments of skins worn by people in very cold countries are made to fit snugly; for not only must the cold be kept out, but the natural warmth of the body must be retained. The human body is like a stove with a fire in it; it constantly generates heat, and in climates where it is very cold it is important to conserve this heat. On the other hand, in very warm countries it is desirable to wear clothing which permits the heat of the body to escape. For this reason loose, flowing garments of linen, silk, or cotton are worn in tropical lands, as the wide trousers of the Turks and Persians, and the unconfined robes worn by other people of Central and Southern Asia.

In countries where it is either very warm or very cold most of the time, the same form of garment is worn year after year. Where the temperature is constantly changing, as in the temperate zone, the style of clothing is also subject to frequent change; and these varying modes constitute what is called "fashion."

While the primary use of clothes is to afford protection from the heat and cold, they should be made and worn with a view to pleasing the eye. It is essential, therefore, that they be carefully cut and neatly made, and they should be kept clean and in good order.

If for no other reason than because so much time and skill are represented in our clothing, we should take good care of it. Each garment we wear represents the work of several wonderful machines and a great deal of skillful labor. There are no more important industries than those which are connected with the making of clothes. In the article on spinning and weaving are illustrations showing some of the machinery which has been invented for weaving cloth. The other articles on the various materials of the sewing room give further data showing how much of the work of the world is devoted to the manufacture of clothing.

COLOR.

Color is an important subject. The author will only attempt to present a few facts in regard to it, which are expressly relevant to the topics treated in this manual.

Beauty in the outer world is of two kinds, harmony of form, and harmony of color. These qualities when combined enhance each other and should always be associated. A perfectly formed garment is far from beautiful if the colors are discordant. The most perfect coloring cannot render a badly proportioned garment attractive. Therefore, although the child may be able, through the system set forth in this book, to cut and make perfectly fitting clothing, if harmony of color is disregarded, her work will be seriously defective. More than this, a study of color is one of the best means for cultivating the perceptive faculties.

Starting with the three primary colors, yellow, red, and blue, the relative value of each should be explained. Yellow makes a quicker impression on the eye than either of the other primary colors. Red is the most perfect color, because it has an equal relation to light and shade. Blue is the most nearly related to shade, and is much slower in reaching the eye than either red or yellow.

The secondary colors, orange, green, and purple, are formed from the primary colors. Orange, which is particularly strong and aggressive, is formed from red and yellow, the two strongest of the primary colors. Green is formed of yellow, which is most closely allied to light, and blue, which is the nearest to shadow, of the three primary colors. It is the most neutral and the softest of the three secondary colors, and, of all decided tints, is the most agreeable to the eye. It is a demonstration of infinite wisdom that the vegetable world is clothed in green; since it counteracts the intense reflection of the sun's rays, and refreshes the eye by its soft and soothing influence. Purple is a union of blue and red, and is a rich and somber color. It was greatly valued by the Romans. A border of purple on their white garments denoted rank. Purple was Cæsar's color. It was made from the Tyrian shellfish, and was really a very ugly hue as compared to the beautiful, rich purple of the present day; but a little touch of it signified so much to the Roman that he valued it highly, and the shellfish of which it was made became an important commercial commodity.

With advanced classes it is desirable to explain the solar or prismatic spectrum, and how its discovery by Sir Isaac Newton established the scientific theory of color. He made the discovery by making an opening, a third of an inch in diameter, in the window shutter of a darkened room, behind which he placed a prism so that a ray of the sun's light might enter and leave it at equal angles. In this way it was found that the ray of light was refracted in an oblong form, and was composed of seven different colors of great brilliancy, — violet, indigo, blue, green, yellow, orange, and red. These colors, when imperceptibly blended together, form what is known as white light.

In arranging color harmony, the first step is to fix on some particular tone or key. If, for instance, a cool green, or gray, or blue which as we have seen is the most quiet and shadowy of the primary colors, is to prevail, the general tone of all the colors must be cool and subdued. If, on the other hand red, orange, brown, yellow, or a warm tint of green be used as the key or prevailing color, the tone of all the colors used with it must be warm. Having decided upon the scheme of color, whether brilliant or subdued, warm or cool, light or dark, let it be remembered that all the beauty of nature's coloring arises from contrast, and that there can be no pleasing combination of tints without variety. Still, the contrasts must not be violent, neither must variety include those combinations which are at variance with the general color scheme or keynote.

In arranging a variety of tints in such a way as to present a pleasing and harmonious whole, there are certain strong colors which must always be used with discretion. This is true of red, which is so positive and obtrusive that it must be very carefully managed and toned. The same is true of yellow, which is much more beautiful in small quantities than in masses. Black, which is the absence of the three primary colors, must also be used with discrimination. It can be used in large quantities only in cool and somber schemes of color. There is really nothing in the whole chromatic series of color more difficult to manage successfully than black and its contrasting hue white. In using black, it should be surrounded and mellowed by deep hues, while white should be introduced by a gradation of the lightest tints ; this, in each instance, prevents a harsh and unpleasant effect. It should be borne in mind that white and black are not colors, but modifiers of color. White stands at the beginning and black at the end of the chromatic scale of colors, but neither the one nor the other is of it.

Some idea of the primary colors should be given the pupils while they are at work upon the first model. If some of the models are done in yellow, others in red, and still others in blue, it will be easy for the teacher to impress upon the children which are the primary colors. In the second model, the three secondary colors may be combined. If it is not possible to get these colors in Saxony yarn, as sometimes happens, the pupils should be taught what the primary and secondary colors are, and should bring to the classroom examples of as many of these colors as possible. Flowers should be brought in their season, that the different color mixtures in them may be studied. It is also desirable to discuss colors in the different fabrics of their clothes, and in such bits of finely colored silk or ribbon as it may be possible to show them.

To lead the children to think about color, and to be interested in its various relations of contrast and harmony as found in nature, is to put them in the way of arriving at correct conclusions. To enable the teacher to do this is all that has been attempted in this brief outline of first principles. While it would be futile, in this connection, to give the rules which govern the numerous differentiations of color, the following includes certain principles which are simple and basic. By uniting two primary colors, the nature of both is altered, and a compound color is the result. As there are but three primary colors in the scale, the two which are united form a contrast to the remaining primary color. Therefore, to reduce the intensity of a primary color, mix with it a certain portion of the color produced by the union of the other two primaries. A simple or primary color thus modified retains, to a certain extent, its nature and characteristic qualities, although subdued and modified sufficiently to render it more capable of harmony with other colors. Illustrations of the results of these combinations may be found in the feathers of birds, in the tints of the human face, eyes, and hair, and in the vegetable kingdom.

SCIENTIFIC SEWING AND GARMENT CUTTING.

CHAPTER I.

OUTFIT FOR SEWING DEPARTMENT.

WHILE it is absolutely necessary that the outfit for a sewing department be complete, it may be very simple and inexpensive. The one described is of this character. It is adequate for a class of from thirty-five to forty children. As a rule, less than an hour twice a week is devoted to sewing, therefore this outfit is sufficient for the accommodation of between one hundred and two hundred pupils.

The low, folding sewing table, with one side laid off in inches and parts of inches, is used as a desk. The cost of these tables is not more than sixty cents each. Four pupils can use one table.

The chairs should be of different heights, in order that the children may all be able to rest their feet on the floor.

The case in which the work and materials are kept (which is illustrated), is simply a series of nine shelves, arranged between two standards four and one-half feet high, placed against the wall. Arranged in tiers of seven on each shelf, are strong pasteboard boxes, furnished

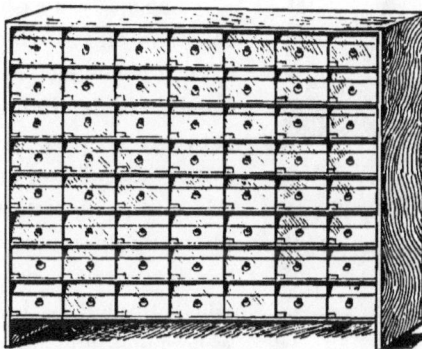

Sewing Case.

with small brass rings, so that they can be drawn out with ease. Each box is twelve inches long by eight wide, and is five inches deep. On the front part, beneath the ring, is pasted a slip of paper bearing the name of the pupil whose work is placed in the box. On the top of this case is a tier of six wooden boxes in which the various wools, threads, strips of canvas, and other small things used in the department, are kept.

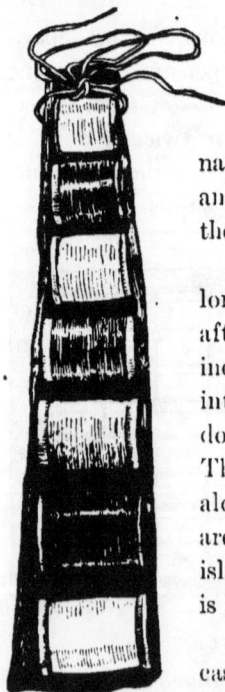

Spool Case.

The little models of the first and second grades are kept in two or three large boxes, the name of the pupil being written on a slip of paper, and pinned to each model. When the pupil reaches the third grade, she is given a separate box for her work.

The scissors case is a piece of cloth sixteen inches long and eleven wide, on which is stitched a strip that, after it is hemmed across its length on one side, is seven inches wide and eighteen long. This piece is divided into twelve parts, and after being basted, is stitched down the width so that twelve little pockets are formed. The fullness which forms these pockets is laid in plaits along the bottom of the case. The bottom and sides are bound with an inch-wide strip which, when finished, forms a half-inch binding. The top of the case is hemmed, and finished with three linen-tape loops.

The spool cases can be made by the teacher. This case is simply a piece of morocco, oilcloth, stiff brown linen, or any substantial material desired ; in size eight and one-half inches long, four inches wide at one end, and three at the other. Cut the edges into seven shallow scallops a trifle smaller at the narrow than at the wider end. Baste in the center of this piece a strip of cardboard eight and one-half inches long, two and one-fourth inches at one end, and one and three-eighths at the other. Cut a silk lining, baste carefully, and bind about the edges with black tape. Put an eyelet in the center of each scallop on both sides ; these may be worked, or metal ones may be used. Double a black silk or linen lacing, and, beginning at the large end with a spool of thirty-six white thread, put it through one eyelet, then the spool, and then the other eyelet, bringing up the sides of the case to the spool. Next put in a spool of No. 40 thread in the same way ; continue to put in each time a finer thread until the case is filled, then tie at the end. The spools revolve on the lacing, and the thread is kept clean, and prevented from tangling.

There should be a swinging blackboard in the sewing room, one side of which is laid off in inch squares, to be used by the teacher in the drafting. Blackboard demonstrations are very essential, and a board laid off in this way makes the objective lessons perfectly clear.

There should be a large table for the teacher's use, and a smooth board fifteen feet long and about two feet wide to lay on two of the small tables to form a cutting table for the pupils. This board has been found to be a most satisfactory arrangement; as it is the right height, the pupils can get around it easily, and it can be laid against the side of the room when it is not in use, and be quite out of the way.

There should be thimbles and needles, tape measures and rulers. The needles should always be of the best. The best thimble for ordinary use is of aluminium, as it is light, does not discolor the finger, and always looks bright and attractive. A large assortment of thimbles should be provided, as in every instance the thimble should fit the finger perfectly.

There are four kinds of canvas used, so arranged that the pupil is gradually brought from doing perfect work on very coarse materials to doing the same work on garment fabrics. The first material is the double-threaded Penelope canvas, which does not strain the unaccustomed eyes of the child. As this is not used in large quantities, a small amount is all that is required.

The next material is Java canvas, also double-threaded, and a trifle

more closely woven than the first, of which more is required, as the model
is larger. The next canvas required is No. 1 Ada canvas, which is used
for the darning. About the same amount of this is needed as of the Java
canvas. More than double the amount of No. 2 Ada canvas is needed
than of No. 1, as the model of this is the largest of the canvas models.

The first garment fabric used is a quarter-inch checked domestic
gingham, either brown or blue. Unbleached cotton cloth is not used
in this course, as it has been found that it is not only the most difficult
material for children to work on successfully, but it soils easily, and, at
best, is unattractive when finished.

The materials used in the advanced grades are good Lonsdale muslin, cambric, coarse and fine linen, and a good quality of flannel. There
should be two cupboards in which to keep these materials and the partly
finished garments. A few yards of cheap calico should be provided to
lay under the materials as they are placed on the shelves, and to bring
up over them, that they may be kept in perfect condition.

A good sewing-machine is a necessity in the advance grades. The
outfit can of course be as expensive and elaborate as is desired; but the
very simple provisions described will serve to indicate what is necessary,
whether it be simple and inexpensive, or elaborate and costly. The ex-
pense of an outfit of course depends wholly on what is selected. The
cost of materials used in a sewing department, which in each instance
must be of the kind and quality called for, is about one dollar per capita
for each grade, averaging the whole course.

While not an absolute necessity, it is most desirable to have a doll
as large as a small child in the outfit of the sewing department. By
having a lay figure of this sort always at hand, the pupils can be taught
to use the system of cutting, when the regular work of the grade is completed, and to draft all sorts of little garments worn by children. More
than this, to make clothes for a big, beautiful doll is always a privilege
highly appreciated; and the prospect of being permitted to make such
garments is to most pupils a strong stimulus to attentive industry.

CHAPTER II.

FIRST GRADE WORK.

PRELIMINARY REMARKS.

THE work of this grade usually occupies between five and six months. It is intended for children of from six to seven years of age, although it has been found equally valuable for beginners in sewing of any age.

As this grade lays the foundation of the entire course, it is most important that everything in connection with it should be very carefully considered. It is especially true in this system of sewing, that the value of forming correct habits in the beginning cannot be overestimated.

The first thing to be impressed upon the children is that their hands must be perfectly clean before beginning to sew, and this point should be carefully looked after by the teacher. The pupils should be seated so that the feet may rest easily on the floor. They should sit erect, with the lower part of the spine against the back of the chair, in such a position that the lungs are not cramped, and that the arms can be used with ease, as in the illustration.

It is important that the pupils of the first grade realize how their work is related to that of more advanced grades. The teacher should call their attention to the illustrations of completed garments, the dressed doll, and other attractive work. They should be assured that when they have learned thoroughly how to do the work of the successive grades, they will be able to make all the garments shown in the models and many besides, and will be competent to make clothes not only for themselves, but for others in the family.

In all kinds of work there are certain rules which must be followed to insure its successful accomplishment. In this case these rules are represented by the squares of the canvas, within which the stitches must be taken in order that they may be perfectly regular, and by the precision required in the slant of the stitches and the drawing of the thread. Since these are basic principles, which, when acquired, enable the children to create that which is excellent, they must be strenuously insisted upon.

As soon as the children have learned how to hold the needle and to take stitches, which is usually accomplished in two lessons, they should be taught to sew buttons onto a piece of cloth of two thicknesses. They should then be held responsible for keeping the buttons on their clothing, and encouraged to perform the same service for the different members of their family. They should be led to have a wholesome pride in neat personal appearance, and the value of their clothes should be impressed upon them by means of interesting facts concerning their texture and manufacture.

THE FIRST MODEL.

The first model is a piece of Penelope canvas five inches long and four inches wide. The double-thread canvas should be used.

When the needles are placed in the pupils' hands, it should be explained that if the hands are not clean the needle becomes rough, and that no one can sew well with a rough needle, or when the material on which it is used is soiled. Explanations of this sort should be made as often as possible in order that the pupils may work intelligently and not mechanically. The first needle used should be a long-eyed and dull-pointed chenille needle. There are three reasons why a beginner should have this sort of a needle: first, because it is a strain on the unaccustomed eyes of a child to attempt to thread a small-eyed needle; second, because often the child cannot thread it without assistance, and

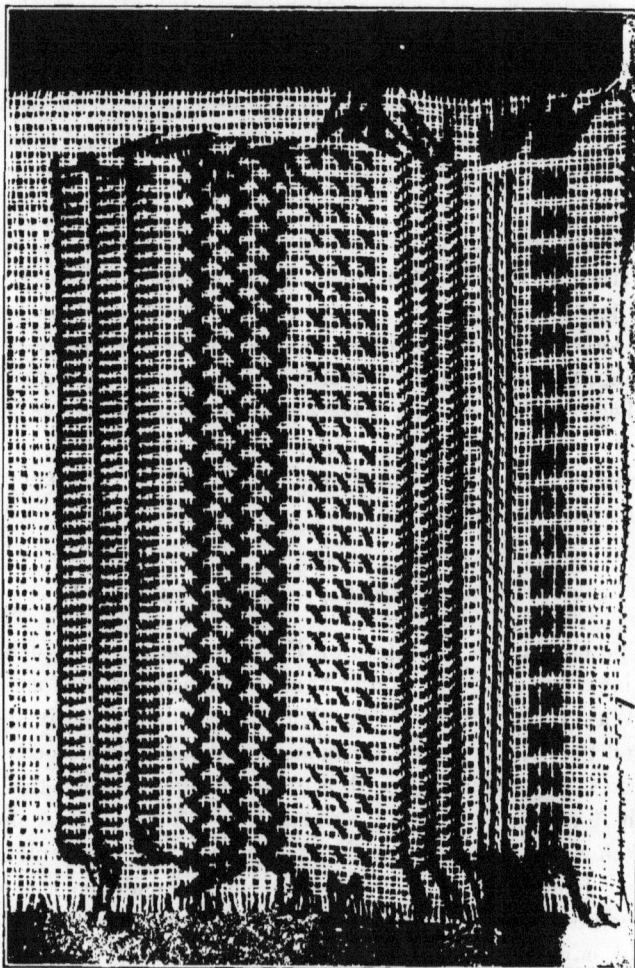

First Model.

it is most desirable to have her work independently from the first; third, because an ordinary sharp-pointed needle is likely to prick the fingers of an untrained worker.

The thimble should be of gold, silver, or aluminium, the latter being the best cheap material for common use.

It should be explained that the thimble is placed on the second finger because it is stronger and longer, and more conveniently situated than the others for pushing the needle through the fabric.

The very best needles should be used, and an emery must be constantly at hand to keep them perfectly smooth.

The thread used in sewing should be just as long as the arm of the one who is using it.

Correct Way of Holding the Needle.

It should be explained that the work is usually done from right to left, and is begun with a backstitch and without knotting the thread. It has been found by repeated experiments that a knot in the end of the thread is not a necessity until the pupil reaches the fourth grade, and it is better that it should not be used until the necessity arises. In this connection let it be remembered that this course of sewing is progressive, and has been arranged in all its details with reference to the general plan of unfoldment as advocated by Froebel and other great educators.

The first thread used is a good quality of cardinal red Saxony yarn; and it should be explained that the proper way of drawing the thread is between the second and third fingers, not only because it is more convenient, but for the reason that it is more graceful.

It is better that the teacher take charge of all implements and the

models used by pupils, until they have passed the second grade, placing them in boxes provided for the purpose.

BASTING.

The first stitch of this system of sewing is the basting stitch. It is begun eight threads from the top and ten threads from the right-hand edge of the model. In putting in this stitch, two threads are taken up and four threads left under the needle at each stitch. There are three lines of this basting across the width of the model, with two threads of the canvas between the lines. Each stitch should be taken with exact precision, and the thread drawn in such a way that the model when finished will lie perfectly smooth. If a mistake is made, in every instance the work must be at once ripped and done anew.

In the very beginning, when the first stitches are taken, the pupils should be taught that the work must be held up towards the eyes, and not the eyes brought down to the work. Insistence upon this and upon sitting erect will insure a correct, hygienic position, which is of the utmost importance. The teacher should explain why these requirements are made.

QUESTIONS AND ANSWERS.

What is the first thing to be done when one is going to sew? *Ans.* To wash the hands very clean and wipe them dry.

Why should this be done? *Ans.* If the hands are not perfectly clean, the needle will become rough and the work soiled.

How should one sit while sewing? *Ans.* With the feet flat on the floor, and the lower part of the body as far back as possible in the chair.

Why is this the proper position? *Ans.* It is easy to sew when sitting in this way, and one does not get tired.

Why is the thimble worn on the second finger? *Ans.* It is the central and the strongest finger, and can push the needle better than any one of the others.

How long should the thread be ? *Ans.* Just as long as the arm.

Should there be a knot in the thread ? *Ans.* No; take a double stitch at the beginning and the thread will not draw out.

What is the first stitch ? *Ans.* The basting stitch.

How is it taken ? *Ans.* Just twice as much is left as is taken on the needle at each stitch.

How should the thread be drawn ? *Ans.* So that it is as tight as, but no tighter than, the threads of the canvas.

Is it necessary that basting be even and the same distance at all points from the edge of the cloth ? *Ans.* Yes; because it is the guide by which a seam is sewed.

How should the thread be drawn in sewing ? *Ans.* Always between the second and third fingers.

THE BACKSTITCH.

The second stitch in the first model is the backstitch. It is well to ask the questions in regard to the conditions of the hands, and the position to be assumed and maintained when sewing, at the beginning of each lesson for six or eight weeks. Thus these most important matters will be so impressed upon the minds of the pupils that correct habits will be formed.

The children should now be taught to sew buttons onto a strip of cloth folded double, and to fasten them neatly and firmly.

QUESTIONS AND ANSWERS.

What is the stitch you are next going to learn ? *Ans.* The backstitch.

Why is it called the backstitch ? *Ans.* The needle is set back each stitch just as much as it is set forward.

How many threads are taken up with each stitch ? *Ans.* Two new threads are taken up, and the needle is set back over the two threads taken up the stitch before.

How should the thread be drawn ? *Ans.* Very carefully, and not too tightly.

How far on the model from the last line of basting is the first row of backstitching begun ? *Ans.* Four threads, and ten threads from the right-hand edge.

When is backstitching used ? *Ans.* When a strong seam is required.

How should all stitches be taken ? *Ans.* Evenly and regularly.

How should buttons be set on ? *Ans.* By sewing through the eyes of the button as many times as the needle will pass through easily ; then fasten firmly on the under side.

OVERHANDING.

First of all, have each pupil double the model together along the third line from the last row of backstitching. That this may be clearly understood, let the teacher fold a model before the class. When this is done, explain that the two sides of the model represent two pieces of cloth. The model having been doubled, let the teacher begin the first row by putting the needle through two threads and leaving two, and continue to carry the thread over at each stitch. When the first line has been completed correctly, show the pupils how to begin the second line, and let them begin the third without help. Between each line of stitching there are two threads of canvas.

QUESTIONS AND ANSWERS.

What is the new stitch you are going to learn called ? *Ans.* The over-hand stitch.

Why is it called the overhand stitch ? *Ans.* Because the thread is put over the edges of the cloth.

For what is overhanding used ? *Ans.* For sewing together the edges of cloth when a perfectly flat seam is desired.

Should the thread be drawn tightly in overhanding ? *Ans.* No; if it is drawn too tightly the seam is not flat, but hard and round.

Should the thread be knotted before beginning to overhand ? *Ans.* No; two stitches, one over the other, are taken to keep the thread from pulling out.

HEMMING.

As in the preceding stitches, after counting a space of six threads of the canvas, begin the first line of hemming by taking a slanting stitch of two threads, leave one space, and take another slanting stitch. Call the attention of the pupils to the neat appearance of the even lines, stitches, and spaces, and let them, as before, start the third line without help. There is nothing more important than a standard of taste, and no opportunity should be neglected to establish a correct standard. This can be best accomplished by inciting admiration for that which is excellent.

QUESTIONS AND ANSWERS.

What is the slanting stitch we are now learning called? *Ans.* Hemming.

For what is hemming used? *Ans.* For sewing a piece of cloth back upon itself.

Why is cloth turned back upon itself? *Ans.* To finish the edge.

Give an illustration. *Ans.* The bottom of an apron.

THE FLANNEL STITCH.

Unlike the four preceding stitches, the flannel stitch is worked from left to right. Although it will be necessary for the teacher to begin the first line, the pupil should now be sufficiently accustomed to counting threads to count off the six threads below the last line of hemming, and show the teacher the point where the first stitch should be taken. After counting off ten threads for the margin and six threads for the space between the last row of hemming and this new stitch, take up two threads from the right to the left. Leave two threads vertically down toward the lower part of the model, and two to the right, and take up two. This makes a diagonal connection between the two stitches slanting toward the right. Leave two threads vertically toward the top of the model and two to the right, and take up two. This again makes a diagonal connection between the two stitches also slanting toward the right. Continue this, and the result is a pretty, vine-like stitch which, although it may seem a trifle difficult at first, can be done, after very little practice, by children from six to eight years of age.

QUESTIONS AND ANSWERS.

What is the fifth stitch on the model called? *Ans.* The flannel stitch.

Why has it been given this name? *Ans.* Because it is mostly used on flannels.

In what way is it different from stitches already done in this model? *Ans.* It is begun at the left-hand side instead of the right, and is worked from left to right.

How much space is left between the lines of the flannel stitch? *Ans.* Four threads of the canvas.

BLANKET STITCH.

Six threads from the flannel stitch and ten threads from the left-hand edge of the model, begin the blanket stitch by taking up on the needle four threads of the canvas vertically, keeping the thread under the needle to form a loop. Two threads to the right of the first stitch take another in every way similar, and so continue across the width of the model. The first stitch should be taken as a backstitch to hold the thread firm, as no knot is used.

QUESTIONS AND ANSWERS.

What is the last stitch on the model called? *Ans.* Blanket stitch.

Why is it given this name? *Ans.* Because it is used for finishing the edge of blankets and other things which are too thick to hem.

How is the blanket stitch begun? *Ans.* At the left-hand side of the model with a backstitch.

How is the stitch taken? *Ans.* Ten threads from the left-hand edge of the model, and six threads from the last row of flannel stitching, take four threads vertically on the needle, and keep the thread under it to form a loop.

What does vertically mean? *Ans.* It means straight up and down.

How many threads are there between each of these stitches? *Ans.* There are two.

THE SECOND MODEL.

For the second model in this grade, cut a piece of canvas eight and a half inches square. It will be observed that this canvas, while simi-

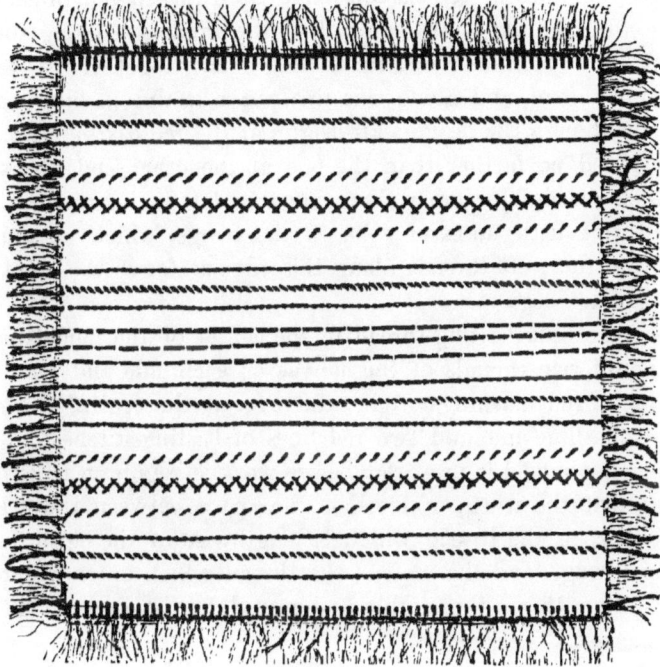

Second Model.

lar to that used for the first model, is still quite different, being softer, more closely woven, and in every way much more like cloth. The reason for this change is that the pupils are thus brought gradually nearer to the various fabrics upon which they will be required later on to do practical work.

When this square has been carefully cut along the line of the threads, from each of the width sides of the model, count ten threads on the length side, or the selvage. Along the line of the eleventh thread put in a row of basting in red marking-cotton. This is the guide for the beginning of the different designs of this second model.

When the lines of red basting have been placed ten threads from what we will call the width edges, fold the square, bringing the sides which have not been marked with basting evenly together. Crease the center by a thread, and along this line put a basting in blue marking-cotton. Although the designs are begun at the red basting at one end, and continued no farther than the one at the other end, the threads with which the designs are worked must be left as long as the model. This will give little notes of color in the fringe, which is formed by raveling out the weft threads along this edge as far as the point where the designs are begun.

When the blue basting-thread in the center of the model has been placed, count two threads of the canvas on each side and put in lines of basting in red marking-cotton, which forms the central design of a group of one blue line and two red lines of basting-stitches.

When this model is prepared, before work is begun on it, explain to the class that the width of cloth is the space between the selvages; that the warp is the thread which is lengthwise of the cloth, in a line with the selvage, and the weft is the thread which extends across the fabric. When this is thoroughly understood, let the work on the model be commenced.

QUESTIONS AND ANSWERS.

Upon what are you now beginning to work ? *Ans.* A new model.

Of what is the new model made ? *Ans.* Of Java canvas.

How large is it? *Ans.* Eight and one-half inches wide by eight and one-half inches long.

What is a figure called that is the same size on all four sides? *Ans.* A square.

How are the threads of all kinds of canvas woven together? *Ans.* In small squares.

What is the first work on this model? *Ans.* To put a line of red basting ten threads from the two weft or width edges of the model.

What are these end lines for? *Ans.* To show where the different designs are to be begun, and where they are to end.

Is the work on this model begun at the side? *Ans.* No; it is commenced in the center and worked both ways.

What is the first design? *Ans.* A blue basting-line in the center of the model with a red one on either side of it.

In beginning the work on these models, is the thread pulled up close to the first stitch? *Ans.* No; it is left long enough to reach to the edge of the model.

Why are the threads used in making the designs left long enough to reach to the end of the model? *Ans.* That the fringe may be made prettier by having a few colored threads mingled with it.

How many colors are used in making the designs of this model? *Ans.* There are two colors, red and blue.

Why are two colors used? *Ans.* To make the model more attractive.

What does this second model form when finished? *Ans.* A very pretty little mat.

THE SECOND DESIGN.

The second design of this model is a union of the backstitch and the overhand stitch. With the exception of the one in the center, each design is repeated on the opposite side from the one on which it is first worked. Let the teacher begin the first design, but in doing so be careful to show the pupil how it is done, telling her that she must begin and complete without help the corresponding design on the opposite side.

For this second design, count three canvas threads from the line of red basting-stitches of the center design, and put in a line of blue back-stitches. Then count two canvas threads, and double the model along the line of the third thread, and put in a line of overhanding in red marking-cotton. Count two threads, and put in a line of blue back-stitching. This finishes the second design. As has already been said, let no inaccuracy pass. When a design is finished, it should be in every particular correct. Anything less than this tends to nullify the educational value of the system, which is intended to be a course of manual training as well as sewing.

QUESTIONS AND ANSWERS.

Of what stitches is the second design composed? *Ans.* Of the back-stitch and the overhand stitch.

What is a composition? *Ans.* It is one thing made up of several things properly put together.

How is this design composed? *Ans.* First there is a row of backstitching in blue marking-cotton, then there is a line of red overhanding, and another line of blue backstitching.

How is the design begun? *Ans.* Count three threads of the canvas from the last red basting-line of the central design, and begin the back-stitching with blue marking-cotton at the red basting-line along the width edge of the model, leaving the thread long enough to reach the edge.

When the first line of backstitching is finished, what is the next thing to be done? *Ans.* Count two threads of canvas from this last line of stitching, double the model together along the line of the third thread, and put in a line of overhanding in red, leaving the thread as long as the model, both where it is begun and finished.

How is this design finished? *Ans.* Count two canvas threads from the overhanding, and put in a line of backstitching in blue marking-cotton.

THIRD DESIGN.

The third design is composed of the flannel stitch and two lines of hemming. In beginning this design, count four threads from the last line of backstitching, and put in a line of hemming in red. Count two threads from this line of hemming, and put in a line of flannel stitching in blue marking-cotton. Again count two threads, and put in a second line of hemming.

QUESTIONS AND ANSWERS.

Of what stitches is the third design composed? *Ans.* It is composed of the flannel stitch and hemming.

How are they arranged? *Ans.* There is a line of flannel stitching in blue with a line of hemming in red on either side.

How far from the last design is the first line of hemming? *Ans.* Four of the canvas threads.

How far from the line of hemming is the flannel stitch? *Ans.* Two canvas threads.

What completes this design? *Ans.* A line of hemming, two canvas threads from the line of flannel stitching.

FINISHING OF THE SECOND MODEL.

The fourth design is a repetition of the second, begun four threads of the canvas from the last line of hemming of the third design. Four threads from this design is a line of blanket stitch extending from one line of the red basting to the other. When each of these designs has been repeated, draw out the red basting-lines which mark the beginning and ending of the designs. Cut the selvage along the line of the first thread, and draw the threads of the four sides, with the exception of the last thread next the design, for the fringed edge. Draw out first the weft threads, leaving the last thread next to the design, on both sides, then draw the warp threads on both sides, leaving the last thread along the line of the blanket stitching, and the model is complete.

QUESTIONS AND ANSWERS.

Of what kind of stitches is the fourth design composed? *Ans.* Of backstitching and overhanding.

How many designs like this are there on this model? *Ans.* There are four.

How many kinds of stitches are there on this model? *Ans.* Six different kinds.

What is this model when it is finished? *Ans.* A little mat ornamented in designs in red and blue marking cotton.

QUESTIONS FOR REVIEW.

What preparation should be made before beginning to sew ?

How should one sit when sewing ?

What stitches make up the designs of the first model ?

How is each of these stitches taken ?

Between which two fingers is the thread drawn in sewing ?

What is the second model of the first grade ?

How many different stitches are there on this model ?

How does the canvas of this model differ from that used for the first ?

What is the warp thread ?

What is the weft ?

What is a square ?

How is canvas woven ?

How many different designs are there in this model, and what colors are used in making them ?

What does this design form when finished ?

Why are the stitches arranged in designs, and why are two colors used ?

NOTE. — The general facts that follow each grade, concerning the more important materials and their manufacture, have been given place, because it has been found that to know something of these subjects stimulates the intelligent interest of pupils in their work. The discussion of these facts, during class work, is optional with the teacher, as they are not a part of the regular course ; but such discussion is recommended. These facts are presented in a condensed form, and it is expected that the teacher will elaborate and adapt them as seems desirable.

The Shepherdess.

Henry Lerolle.

MATERIALS AND THEIR MANUFACTURE

WOOL.

THERE was once a little white lamb, with mild eyes and a short woolly tail, that lived near the Pacific Ocean in a pretty green valley with high mountains on either side.

When this little lamb was about four weeks old, it began to nibble grass and other green things: before this its only food was its mother's milk. It grew a little every day, and when it was eight months old it was almost as large as its mother.

One day the Mexican shepherd who took care of this lamb, its mother, and three or four hundred other sheep and lambs, took it to a little stream that ran through the valley, and washed it. Then, after putting it in a pen, he cut off all its soft white wool. The shepherd was so skillful that, when he had finished cutting, the whole fleece was in a single sheet the size and shape of the lamb. The fleece is the wool of a sheep or lamb after it has been cut from its body.

When the shepherd had clipped the wool from each of the lambs that were eight months old, he packed all the fleeces together in great sacks. These sacks were sent to a place called a factory. The first wool cut from a lamb is the best; but there are different kinds of wool, even in a single fleece. Some parts of these fleeces, therefore, were made into very fine, soft cloth, and some into the nicest lamb's-wool yarn, and such delicate wool thread as is used in working the models in the first and second grades.

The lamb we are reading about was of the variety known as the Spanish merino; and like all sheep of this kind, it had rather a small body, and very long, thick wool. When its wool was cut off, it could walk and run much easier than before. It was then taken with the other sheep to a State a long distance from where it was born, called Wyoming. The farmers there wanted some long-wooled Spanish merino sheep; for they had only South-Downs, Leicesters, and other common varieties.

The Mexican shepherd, dressed in an oilskin coat and trousers, with a blue shirt of sheep's wool trimmed with gay buttons and lacings, and a broad-brimmed hat, went with our lamb and the other sheep, to take care of them. Sheep must always have some one to look after them; for they are gentle, helpless creatures, and it matters not how old they are, they never seem to know enough to take good care of themselves. But though in some ways they require more care than other animals, man can well afford to give them attention, for they are very useful indeed. Their wool, which is a species of hair, is one of the most valuable materials in the world for all kinds of clothing. Their flesh, which is called mutton, makes very good food; and their skins are made into a leather that is used for many purposes.

If our lamb had lived in some other part of the world, it would have had a very different kind of a shepherd to care for it. In Scotland he would have been, in olden times, a blue-eyed, light-haired man, with a long white cloak made of the locks of the sheep. He would have carried a crook, or crosier, as a staff; a sling for throwing stones; and a pipe or flute on which to play while his flock ate grass. With him there would have been a dog, trained to help drive and care for the sheep.

In Yorkshire, England, in the olden times, a shepherd was quite an important man, who had a great many friends; for there were many shepherds in Yorkshire. They had one Sunday in the year set apart for them, called the "Shepherds' Sunday;" and the time when they cut the wool from their sheep, called "sheep-shearing time," was observed as a great festival.

If our lamb had lived in France, he would very likely have had a young girl to take care of him and the rest of the flock. The French shepherdess knits while she tends her flock, instead of playing on a pipe like the Scotch shepherd.

It was a long journey from the first home of our lamb to Wyoming; and as the sheep walked all the way, it took them a good many days to get there. But the longest way is finally passed if we keep steadily on, and the hardest task is at last accomplished if we do a little every day. In some of the places through which the sheep passed, there was very little water and almost no grass. Such a place is called a desert. The way was not only long, but hard, and the sheep and lambs often lay down to rest. Sometimes the Mexican shepherd also got so heated and tired that he dug away the hot top of the earth; and when he came to that which was cool, he put his oilskin coat over some low sage bushes, in order to make a little shelter from the sun, and lay down for a time.

At last the lamb, its shepherd, and the other sheep, arrived at their new home. I wish you could have seen it! There were bright flowers, green grass, blue skies, a pretty brook that emptied its water into a river not far away, and many other pleasant and beautiful things. By this time the wool had grown again all over the lambs; but the shepherd did not cut it off now, for winter was at hand, and they would need this thick covering to protect them from the cold winds and the snow. When the spring came once more, bringing warmth and sunshine and the flowers, the shepherd cut the wool from all his sheep; for they could then do without their heavy coats, just as boys and girls can wear lighter clothing when it is warm. The lamb had now grown to be a young sheep; and although it had more wool than when it was first sheared, its fleece was not worth so much, because it was not lamb's wool. Still it was very nice, because it was of the merino variety; and the merino sheep, even when they are old, have fine wool.

CHAPTER III.

SECOND GRADE WORK.

PROVIDED the pupil devotes forty minutes twice a week to sewing, the work in this grade will occupy a school year.

The one model for this grade is a canvas bag ornamented with simple designs, which, when properly completed, exemplifies all the stitches taught in the course. When the pupil has finished it, she is ready to work on garment fabrics.

The aim of this system is to make it possible for the pupils to do the work of each lesson, not without instruction, but without assistance. It has been demonstrated that this is possible in every case when each step is thoroughly understood as the pupils proceed. The preparatory work cannot be too carefully considered, as with all first principles; and although a year seems a long time to work on a single model, it will be found none too long for the pupils to become quite familiar with the stitches which in the succeeding grades are used under more difficult and exacting conditions.

The children, having now become familiar with the use of the needle, should be encouraged to work independently outside the classroom. Towels for kitchen use in the home may be hemmed and brought to the teacher for inspection. The combinations of the various stitches, as they are learned, can be used in ornamenting dolls' clothes, flannel petticoats, and little spreads for tables or washstands. An excellent practice is to let the children bring in designs made outside the class, by combining, according to their own ideas, the various stitches they have learned. In this way the creative faculty is stimulated, and at the same time the work is correctly done.

THE MODEL.

The material of this model is No. 1 Ada canvas, which, while soft and like cloth in many ways, is yet so coarse that the threads can be easily counted. It is eighteen inches long and nine inches wide; and when it has been carefully cut along the line of a thread, it may be given to the pupil to overcast. The overcasting should be explained as an overhand stitch, longer and deeper than the one used in sewing. The pupils should be required to do this work neatly, taking the stitches four threads apart and four threads down. It must be explained that this overcasting is necessary, since otherwise the goods would ravel or fray along the edges.

Canvas Bag.

When the model has been overcast, measure two inches from the top and bottom, draw a thread, and put a line of red basting in the space. This is to indicate where the designs begin and end.

QUESTIONS AND ANSWERS.

What will the new model be when it is finished? *Ans.* A fancy bag.

Of what is the model made? *Ans.* Of canvas.

What is the size of the model? *Ans.* It is eighteen inches long and nine inches wide.

What is the first thing to be done on this model? *Ans.* To overcast the edges.

How is overcasting done? *Ans.* Like overhanding, only that the stitches are deeper, and farther apart.

What would happen if the edges were not overcast? *Ans.* They would fray.

THE FIRST DESIGN.

Like the last model of the first grade, the designs of this are worked from the center.   First let the pupil put the long edges of the model carefully together, and crease the center.   From the center thread count four to the right, and put in a row of backstitching.   Let this backstitching be between the fourth and fifth threads from the center.

The thread used is red crochet cotton, which makes a better design than marking-cotton, and is less expensive.

When this line is completed, count nine threads to the left, and put in another row of red backstitching.   In the center between these two put in a line of red flannel stitching, worked over three threads of the canvas.

QUESTIONS AND ANSWERS.

In what part of the model is the first design worked? *Ans.* In the center.

How is the center of the model formed? *Ans.* By placing the two long sides of the model together, and creasing it along the center.

What is the first stitching put in? *Ans.* A line of backstitching between the fourth and fifth threads from the center.

What is the next step? *Ans.* Nine threads to the left of the first line of backstitching put in another line.

How is this central design finished? *Ans.* With a line of flannel stitching between these two lines of backstitching, over the three center threads of the model.

In what color and kind of thread are all the designs of this model worked? *Ans.* In red crochet cotton.

THE SECOND DESIGN.

Between the first and second designs, there is a space of six threads of the canvas. Count seven threads to the left; crease the model between the seventh and eighth threads, and put in a line of overhanding by taking up two threads and leaving two, forming a diagonal bar across a square of four threads. When this is completed, count eleven threads to the left, crease the model between the eleventh and twelfth threads, and put in another line of overhanding.

When these two lines are completed, there are ten threads of canvas between them. Begin the line of flannel stitching that finishes this design three threads from the last line of overhanding, and carry the thread in the needle over four canvas threads; then proceed as in the flannel stitching already described, keeping four threads between the upper and lower stitches, instead of three as in the central design.

QUESTIONS AND ANSWERS.

How many threads are there between the first and second designs? *Ans.* There are six.

What is the first stitch of the second design? *Ans.* The overhand stitch.

Where is the model doubled down for this stitch? *Ans.* Between the seventh and eighth threads of the canvas, counting from the last line of backstitching of the second design.

How many threads are there between this and the second line of overhanding? *Ans.* Ten threads.

Where is the canvas creased for the second line of overhanding? *Ans.* Between the eleventh and twelfth threads to the left of the first line.

How is this design finished? *Ans.* With a line of flannel stitching between the two lines of overhanding.

Where is this placed? *Ans.* The first stitch is placed three threads from the last line of overhanding, and the next across four threads, and three threads from the other line of overhanding.

THE THIRD DESIGN.

For the third design, leave a space of six threads of canvas, and between the sixth and seventh threads put in a line of backstitching. Count three threads, and put in a line of basting, taking up two threads and leaving four. Count two threads, and put in another line in the same way, and again count two threads, and put in a third line. Leave a space of three threads, and put in a line of backstitching, which finishes the third design.

QUESTIONS AND ANSWERS.

How far is the third design from the second? *Ans.* Six threads.

Of what is it composed? *Ans.* The backstitch and basting-stitch.

Where is the first line of backstitching placed? *Ans.* Between the sixth and seventh threads from the second design.

Where is the first line of basting placed? *Ans.* Three threads from the line of backstitching.

How far from the first line of basting is the second one placed? *Ans.* Two threads, and the third and last is two threads from that.

How is this design finished? *Ans.* By a line of backstitching three threads from the last line of basting.

THE FOURTH DESIGN.

Leaving a space of six threads, put in a line of hemming by taking up two stitches and slanting over two. Count three threads, and put in a line of stem, or, as it is often called, outline stitch. This is done by taking up two stitches and going back two, beginning at the left instead of the right, and keeping the thread from the beginning to the end on the same side of the stitching. This stitch is really the reverse side of backstitching, and on the under side of the model forms a perfect backstitch.

There are three lines of the stemstitch with two threads of canvas between each of them. Three threads from the last line of stemstitching put in a line of hemming. This finishes the last design on the left side of the bag. Each of these designs is repeated on the right side.

How many designs ornament this bag? *Ans.* Seven in all.

How many different designs are there? *Ans.* Four.

How many designs are repeated? *Ans.* Three are repeated.

Of what is the last design composed? *Ans.* Of hemming and the stemstitch.

How are they arranged? *Ans.* First there is a line of hemming six threads to the left of the backstitching that finishes the last design.

What is the next stitch in this design, and where is it placed? *Ans.* The stemstitch, which is put in three threads from the hemming.

How is the stemstitch put in? *Ans.* It is begun at the left instead of the right, by first taking up two threads, then setting the needle back two and taking two threads at each stitch, keeping the thread under the needle.

What stitch is the stemstitch like? *Ans.* It is taken like, and is really the reverse side of, the backstitch.

How many lines of stemstitch are there in this design? *Ans.* Three; and there are two threads of the canvas between them.

What finishes this design? *Ans.* A line of hemming three threads from the last row of stemstitching.

HEMSTITCHING.

The designs being finished, the next work is the top of the model. It will be remembered that a thread of canvas was drawn, and a red basting placed to mark the limit of the designs. Draw out this red thread and one more thread of canvas, and the model is ready to be hemstitched across the ends, which later on form the top of the bag.

As has been said, each step in this course of work has been arranged, after repeated experiments and long experience with children of different ages, in accordance with the principles of a harmonious development rather than a forcing of faculties. It has been found that it is too difficult for beginners to undertake to catch the hem and learn the hemstitch at the same time, therefore there is no hem turned in this first lesson in hemstitching.

When the threads of canvas have been drawn, let the pupil first take a backstitch to secure the thread, then take up two of the warp canvas threads which have been left by pulling out the weft, catching it into one thread of the cloth below. Set the needle back two threads, and take up two threads, as in the backstitch, setting the needle each time one thread into the firm cloth.

When the ends have been hemstitched, measure down the length of the model twelve threads from the last design, and draw two threads on each side. Turn in the edge of the model one-fourth of an inch, crease down a hem that just reaches to the drawn threads, and baste.

When this has been done, let it be hemstitched, being careful to explain that not only must two threads be taken up at each stitch, but also the edge of the hem, as in this way it is held in place.

QUESTIONS AND ANSWERS.

What is the first thing to be done in hemstitching? *Ans.* To draw the threads.

How is the stitch taken? *Ans.* Two threads are taken up and the needle set back over two at every stitch.

How is the hem turned? *Ans.* First the edge is turned one-fourth of an inch, and the hem is then creased down so that the edge just reaches the drawn lines.

Was there a hem turned at the top of the bag? *Ans.* No; that was a practice hemstitch.

How is the hemstitching along the edge of the bag different from that at the top? *Ans.* At the side, the edge of the hem is caught with every stitch.

Is hemstitching useful? *Ans.* Yes; in ornamental work.

Should hemstitching be done very evenly? *Ans.* All stitches should be done evenly.

JOINING AND FINISHING THE MODEL.

Fold the model with the long sides together so that the two short sides are even, and overhand the hemstitched sides, carefully matching the threads. The attention of the pupil should be especially called to the depth of the stitch. It should not be more than one thread deep, as otherwise the seam will be clumsy, and will not lay flat. No. 40 white cotton thread should be used for this overhanding.

After turning down the top of the bag five threads, baste a red tape half an inch wide about the top, one thread from the edge, so that it does not show on the right side. It should be basted twice, once near the upper and once near the lower edge. The lower edge of this facing must be even with a thread of the canvas, where it is held in place by a line of hemming in the red crochet cotton.

The upper edge is finished with the buttonhole stitch in red crochet cotton. The stitch is made by setting the needle down four threads, and throwing the thread from the eye of the needle over it from right to left, forming a twisted loop, which is the pearl edge of the buttonhole.

When the top is finished, draw six threads above the horizontal line of hemstitching about the top of the model, and from each side run in a piece of No. 2 red lute-string ribbon, which is long enough to form a bow on either side when the bag is not drawn up. This completes the work of the second grade.

QUESTIONS AND ANSWERS.

How is the model finished? *Ans.* It is first folded so that the short edges come evenly together.

How is it joined? *Ans.* It is overhanded together very carefully, bringing the hemstitching at the top and the threads down the sides evenly together.

How deep should the overhand stitch be taken? *Ans.* One thread on each side.

What thread is used? *Ans.* No. 40 white cotton thread.

What is the next thing to be done? *Ans.* Face the bag with red tape a half an inch wide.

How is this done? *Ans.* Turn in the top of the bag three threads, baste the tape on one thread from the top, having the lower edge straight with a thread, where it is hemmed on with red crochet cotton.

How is the bag finished at the top? *Ans.* With the buttonhole stitch in red crochet cotton.

How is this buttonhole stitch done? *Ans.* The needle is set four threads down, and the thread looped over the needle from right to left in such a way as to form a pearl or buttonhole edge.

What is the next thing to be done in finishing this model? *Ans.* Above the hemstitching, around the top of the bag, draw six threads.

What is this space for? *Ans.* For the two ribbons used in drawing up the bag.

How are these ribbons run in this space? *Ans.* Under six threads and over six threads.

Why are they run from both sides? *Ans.* So that the bag may be drawn evenly from each direction.

What sort of ribbon is used, and how is it finished at the ends? *Ans.* No. 2 red lute-string ribbon is used, and it is tied in a bow on each side.

How much ribbon is required for this bag? *Ans.* Two yards.

QUESTIONS FOR REVIEW.

How many models in the second grade?

What is this one model?

In what way is the canvas of this model different from that used for the second model of the first grade?

How many different designs ornament this model?

What is a design?

What is a composition?

How many different kinds of stitches are there on this model?

How many of these stitches have not been given before?

Describe them.

When is overhanding used?

Of how many stitches is each design composed?

Are there more stitches used in ordinary sewing than there are on this fancy bag?

What is the difference between a flannel and a buttonhole stitch?

How does stemstitching differ from backstitching?

MATERIALS AND THEIR MANUFACTURE

FLAX.

A COTTON field, with its opening pods, or, as they are called where cotton is raised, bolls, is a beautiful sight; so also is a field of blooming flax. The one is like a sea of gleaming silver, the other like a sea that is as blue as the sky.

The blossom of the flax plant is a delicate and beautiful shade of blue. Unlike cotton, flax, from the fiber of which linen is made, grows best where it is cool. When it blooms, the plant is between two and three feet high. It requires a great deal of moisture, and it is therefore most successfully cultivated in the lowlands of Holland and Belgium.

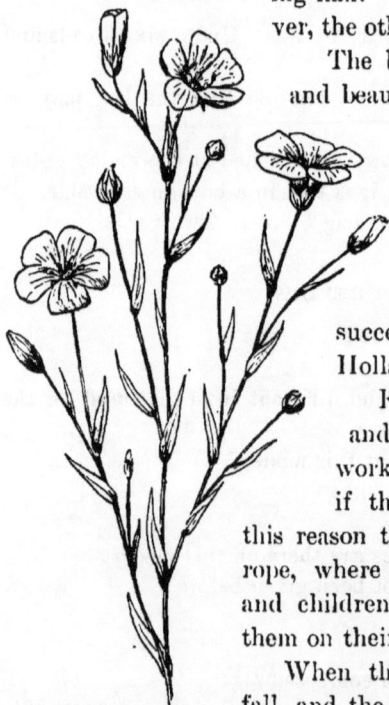

Flax is, so to speak, a delicate plant, and it therefore requires a great deal of work to raise it. It will not grow well if there are any weeds near it, and for this reason they must all be pulled up. In Europe, where the best flax is cultivated, women and children weed the flax fields, going through them on their hands and knees.

When the leaves of the flax plant begin to fall, and the stock to turn yellow, it is harvested.

Flax.

This is done by pulling the plants up by the roots, and laying them evenly together, as the fiber of which the linen is made is injured if they are twisted or doubled. This fiber lies between the bark and the inner, woody pith of the plant; and it is rather a long and tedious process to separate it.

When the flax has been pulled, the first thing done is what is called " rippling " it, which is removing the seed-pods. The next thing to be done is the " retting," which is a fermentation that loosens the gummy substance which binds the fiber to the wood. This is accomplished by exposing the flax to the dew in the fields, or by immersing it in water.

Hackling Flax.

To put it into water is better than to depend on the dew; in fact, it is the only way to get really fine fiber. The flax stalks are kept wholly under water, but are not permitted to rest on the bottom of the pond or tank.

This retting process, which separates the fiber from the rest of the plant, requires both skill and care. If the stalks are left too long in the water after fermentation has taken place, the fiber is weak and lacks gloss. If it is not left long enough, it is dry and coarse. Again, the water used must be pure, soft, and free from lime, iron, or other substances of a similar nature which color and injure the fiber.

The water of the river Lys in Belgium is expressly suited to retting flax, and for this reason the flax grown near it is the finest in the world.

The next thing after the retting, is to remove the woody pith. This is called "scutching," and is accomplished by beating the flax until the wood drops out and the fiber is left. Sometimes this is done by machinery, and sometimes by hand.

After all this has been done, the flax is "hackled," and it is then ready to spin. The hackling is a combing process by which the chaff and the coarse, rough fiber, called the tow, is removed, and only the clean linen fiber is left, white, straight, and ready for the spinner.

Sometimes flax is cultivated, not for the fiber, but for the seed, which is used for making linseed oil. It is then permitted to get thoroughly ripe, much riper than when the fiber is to be used. The seed only is gathered then, and the stalks are thrown away.

So long ago that there is no written account of it, flax was cultivated for the fiber, which was used, as it is now, for clothing. We know this because pieces of linen have been found in tombs and other places where it had been lying thousands of years. Although the art of making linen from flax is so old, there has been very little change in the way in which it is prepared for spinning; and the process is much the same as it was when the children of Israel were in bondage in Egypt. Until very recently it has been almost wholly a domestic art. Even now there are small farmers in Scotland and Ireland who raise flax; and it is prepared for spinning, spun, woven, bleached, finished, and made ready for market, by the farmer's wife and children.

THIMBLES.

Just fancy how awkward it would be to wear a thimble on your thumb. Yet for a good many years after thimbles were invented they were worn only on the thumb. Because of this they were called thumb-bells. After a time this word was shortened, and the very useful little contrivance with which a needle is pushed through fabrics was called, as it still is, a thimble.

When the thimble had been in use for some time, it was found that it could be used much more successfully on the middle finger than on the thumb; and now it seems strange that it should ever have been used in any other way.

The thimble was invented in Holland. It was in 1695 that John Loftington came over from Holland, and established a manufactory of thimbles at Islington, England. At that time, and for a long time afterwards, thimbles were made entirely by hand; and many of them were beautifully wrought, and set with gems. Now all this is changed, and with very few exceptions they are made entirely by machinery.

The ordinary thimble, whether of gold, silver, steel, aluminium, celluloid, or any other material of which thimbles are made, is first molded into the size and form desired. The small indentations in which the eye of the needle rests as it is pushed through the fabric are made by machinery. The polishing is also done very rapidly by machinery, all of which is simple, and needs very little attention. Hence the labor required in making a thimble is small; and thimbles cost but little, save when they are made of expensive material.

CHAPTER IV.

THIRD GRADE WORK.

UP to this point the pupil has been engaged in becoming familiar with the needle and thimble and the different stitches used in sewing. Now the scissors are added to the implements which will be constantly employed, for the work of this grade includes some of the fundamental principles of drafting and cutting. In this work, as in the sewing, the natural method of development by gradual unfoldment is followed.

The first thing is to teach the child to use the scissors. Scissors about five inches long should be selected; and they should be of good steel, and sharp. The first material used should be heavy manilla drafting-paper, laid off in half-inch squares; and the pupils should be required to cut carefully along the lines until they can follow them accurately.

When this has been accomplished, they may be given a piece of checked gingham to cut along the line of the design. When the pupils can cut a straight line, they may draft the first model of this grade.

First draw a parallelogram twenty-four inches long and twelve inches wide, on the blackboard, explaining what a parallelogram is. Then let the pupils draw and cut out of the manilla drafting-paper a parallelogram half this size, which can be done easily, as the paper is laid off in half-inch squares. After this let each pupil cut a parallelogram of the gingham twenty-four inches long and twelve inches wide. This will not be difficult, the straight lines of the checked design forming an accurate guide. This is the simple outline of the gingham case, which is the first model in fine thread fabric.

When the parallelogram has been cut, measure at one end three inches each way from both corners, draw oblique lines, and cut off the corners. From the lower corners of the oblique lines, measure down eleven inches, and draw a dotted line to show where the parallelogram is folded to form the case.

Turn in the edge along the last thread of a white line of squares, crease evenly through the center of the next line of brown squares, and baste so that the design is not broken. Hem on the last white thread of the third line of squares, beginning to count after the first edge has been turned in. Let the hemming be done on this last white thread of the third square. This may seem an unnecessary exaction; but let it be remembered that the educational value of this system is lost if the nicest precision is not observed. Again, when correct habits are formed, it is as easy — nay, it is easier — to do a thing well as ill.

For this hemming, No. 70 white cotton thread should be used. The stitch should be so lightly taken that it is not noticeable on the right side and along the first white thread of the last check.

Model of Gingham Case.

Turn and baste the straight and diagonal ends of the model the same as the sides, being careful to exactly match the design of the model to a thread. When these hems are basted, turn them back, and

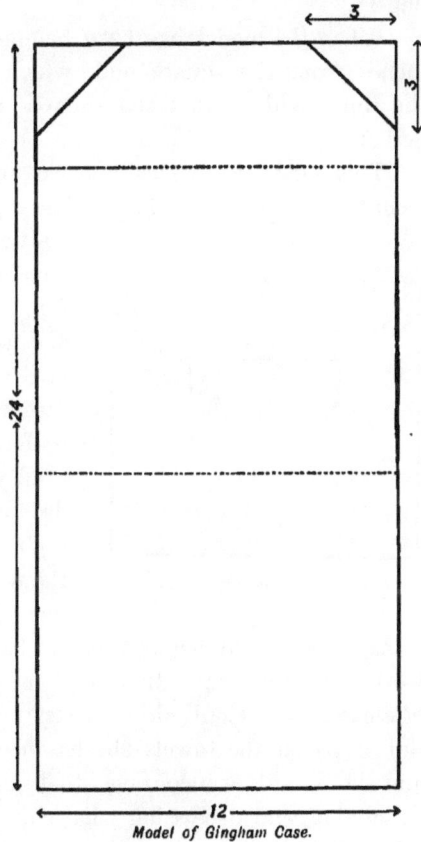

overhand them with fine, shallow stitches, instead of hemming in the usual way. This is called the linen hem, as it is used for table napery.

When the model has been hemmed, fold it on the dotted line, ten inches from the square end, which leaves four inches for a flap at the top ; hold so that the squares match exactly, and overhand together.

This model is finished by two linen-tape loops, sewed on one inch from the corners of the flap. The tape is first overhanded together the width of the hem, and then is overhanded onto the hem and across the top. The flap is then turned down ; and the buttons, which should be covered with white linen, are placed in the center of each loop, with a piece of tape under each to hold it firm.

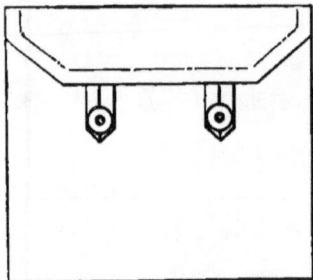

By the work of this grade the children are made capable of doing at least a portion of that most important part of the work of the household, — the mending. The teacher should make it a point to have each child, as soon as she has learned to do a certain kind of work well, undertake that work at home, so as to assist her mother. She can now not only do a variety of stitches and darning, but she can put loops on the towels she has hemmed, and assist in plain sewing ; and she should be encouraged to do this work independently, the teacher examining and discussing what she does at home. She should also be required to keep her own clothes mended. While faults must of course be recognized and pointed out, severe criticism of work done independently should be avoided ; and merit should be praised, in order that the children may not be discouraged, but be incited to unaided and original effort.

Model of Gingham Case Finished.

What is drafting? *Ans.* It is drawing a plan or pattern.

What is the first thing to be done when one is going to draft? *Ans.* To take the measures and set them down.

What is a parallelogram? *Ans.* A figure bounded by four straight lines with opposite sides parallel.

What is the shape of the first model of this grade as it is drafted? *Ans.* A parallelogram.

What are the dimensions? *Ans.* Twenty-four inches long by twelve inches wide.

When the measures have been taken, and the paper parallelogram pattern drawn and cut, what is the next thing to be done? *Ans.* Pin on the cloth, and cut very carefully on a thread of the gingham.

What will this model be when it is finished? *Ans.* A case.

How is the flap formed? *Ans.* At one end measure three inches each way from both corners, draw oblique lines, and cut off the corners by them.

What is the next thing to be done after the case is cut? *Ans.* Measure down eleven inches, and draw a dotted line to show where the parallelogram is folded to form the case.

When this is done, what is the next step? *Ans.* Turn in the edge, and crease a hem along the second line of the squares and baste so that they match exactly.

How are the two ends of the model finished? *Ans.* They are hemmed in the same way as the sides.

How is this hemming done? *Ans* With the overhand stitch.

What is this sort of hemming called? *Ans.* Linen hemming, because it is used for table linen.

How is the case formed? *Ans.* By doubling it along the line drawn to indicate the bottom of the case, bringing the squares of the design together so that they match exactly, then basting and overhanding together.

How is this model finished? *Ans.* With two linen loops, one on each side, overhanded together, the width of the hem hemmed on with an overhand stitch, and two buttons sewed on, with a piece of tape under them to make them strong.

SECOND MODEL.

The second model in this grade is a piece of canvas, six and one-half inches square, upon which the first two kinds of darning of the course are done. Let the pupil measure, draft, and cut this square, which, although small and simple, is somewhat more difficult to draft and cut than the gingham with its clearly defined designs.

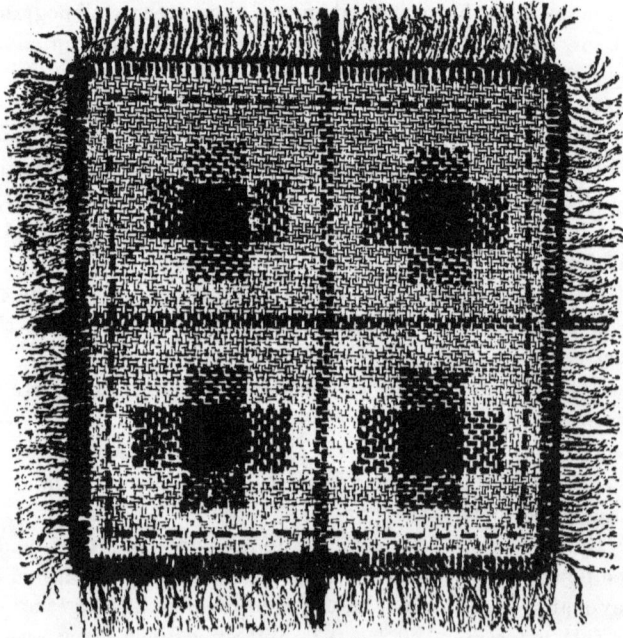

Canvas Darning Square.

When it is cut, let it first of all be overcast. Then put the edges together and crease through the center, and run a thread of red Saxony wool either side of this crease, taking up one thread and leaving one with a thread of canvas between the two. Fold the canvas in the opposite direction, crease through the center, and again run two threads of red Saxony wool on either side of this crease, leaving a thread of canvas between them, and dividing the model into four squares.

From these lines count thirty-four threads of canvas each way, and put in a line in the red wool, taking up two threads and leaving two around the whole model.

Three threads from the line which forms a square within the square put in a solid edge line of blanket stitch in red Saxony wool on each side of the model. The fringing of the model outside of this blanket stitching is left until the last, so that when it is finished it may be quite fresh and clean.

To prepare the model for the two sorts of darns which it exemplifies, count off in each of the four small squares twelve threads from the side and six threads from the top and bottom. Begin at the top, and run an outline thread to the point six threads from the bottom. Count ten threads to the right, and run another thread like the first, beginning six threads from the top, and ending the same distance from the bottom of the small square. This forms a bar ten threads wide and twenty-two long. Outline another bar crossing this at right angles in the same way, and the same number of threads wide and long. After outlining two bars of this sort in each of the four squares, baste a piece of cardboard under one of them.

The two outlined bars under which the cardboard has been placed form a small square in the center of a square. Six threads beyond the limit of this square, along the line of the outline bar, the darning is begun by taking up one thread and leaving one, until the center square, formed by the two bars, is reached. At this point leave ten threads, drawing the red Saxony darning wool straight across, and again take up every other stitch on the other side of the square for six stitches. Continue this until the opposite outline of the bar is reached. Then turn the model, and fill in the bar that crosses this at right angles in the same way, with the exception that the loose center warp threads are woven under and over, each alternate thread being taken up on the needle. In the opposite diagonal corner repeat this darn, which is known as the stocking darn.

The darning in the third square of the model is diagonal or linen darning. The first threads, which represent the warp, are straight across, like those in stocking darning. Outside of the small central square, the threads are woven over and under as in stocking darning. When the center is reached, take up two threads, and leave two the first time across. The second time across, first take up a single thread, and after that take up two and leave two. The third time across, first leave two, then take up two and leave two. The fourth time across, leave the first thread, and after that take up two threads and leave two. Repeat this, beginning with the first, until the square is filled. The fourth square is done in the same way. The practical application of this darning is to baste a piece of cardboard under the hole, which is then cut out square. If it is linen or any diagonal weave, use the linen darn; and if the under and over weave, use the stocking darn.

QUESTIONS AND ANSWERS.

What is the second model in this grade ? *Ans.* A square of canvas six and a half inches each way.

What is the first thing to be done after drafting and cutting this square ? *Ans.* To overcast the edges.

Then what should be done ? *Ans.* Put the two edges together, crease through the center, and run a thread of red Saxony either side of this crease, taking up two threads and leaving two.

What is the next step ? *Ans.* Fold the model in the same way in the opposite direction, crease, and run a thread of red Saxony either side of the center.

What do these two lines put in from opposite sides form ? *Ans.* Four squares within the model.

What is the next step ? *Ans.* Count thirty-four threads each way from these double lines, and at this distance run a line of red Saxony on each side of the model.

What is next to be done ? *Ans.* Three threads from this line of running that bounds the square, put in on each side a solid line of blanket stitching in red Saxony.

How many kinds of darns are there on this model ? *Ans.* Two, the stocking and the linen darn.

How is the stocking darn begun ? *Ans.* Count off in one of the four small squares twelve threads from the sides and six at the top and bottom. Begin at the top, and run an outline thread to the point six threads from the bottom.

Where is the next thread placed ? *Ans.* Count ten threads to the right, and run another thread like the first, which forms a bar ten threads wide.

What is the next step ? *Ans.* Outline another bar, in every way like this, at right angles with it.

Is this kind of crossed bars outlined in each of the four squares ? *Ans.* It is.

What should always be basted under material that is to be darned ? *Ans.* A piece of cardboard.

After the cardboard is basted under the crossed bars, how is the darning done ? *Ans.* Six threads toward the edge from the square formed by the crossed bars, begin the darning next to the outline thread, taking up one thread, and leaving one until the center square is reached.

What does the center square represent ? *Ans.* The space to be darned ; the thread is taken over it without stitches.

Where do the stitches begin again ? *Ans.* On the other side of the square ; one thread is taken up, and the other left for six threads.

What do these threads represent ? *Ans.* The warp.

How are the cross threads or weft of stocking darning put in ? *Ans.* Like the warp thread, except in the center, where it is woven under and over the warp threads.

How many times is this darn repeated in model ? *Ans.* Once.

In what part of the model is the first linen darn ? *Ans.* In the third square.

How is it put in ? *Ans.* The warp is put in like the stocking darn.

How is the first thread of the weft put in ? *Ans.* Until the center of the cross bar square is reached, the thread is put in over and under the canvas threads, the same as the stocking darning. At the center, take up two threads and leave two.

How is the second thread put in ? *Ans.* Like the first until the center of the bar is reached, then take up one thread, and after that leave two and take up two.

How is the third thread put in ? Like the first and second until the center is reached, then begin by leaving two, take up two and leave two.

How is the fourth thread put in ? *Ans.* The same as the others as far as the center, then, leaving the first thread, take up two and leave two.

How is this darning finished ? *Ans.* The way in which the first, second, third, and fourth threads are taken up are repeated until the square is filled.

How is this model finished ? *Ans.* By raveling out the edge to the blanket stitching.

How are these darns used in mending ? *Ans.* A piece of pasteboard is basted under the hole, which is then cut square. If the fabric to be darned is over and under wove, the stocking darn is used, if diagonal, the linen darn.

THIRD MODEL.

For the third model in this grade, which is for the knitted darning, cut a piece of cardboard three and one-half inches long and two and one-half inches wide. Draw straight lines one-half an inch from the top and bottom. This should be done by the pupil without assistance, with directions from the teacher. The lines at the top and bottom of the card should be divided into eighth-inch spaces. When this is done, take red cotton thread and stitch in each division, drawing the thread from the upper to the lower line, and so setting the needle that the thread is not straight, but slants from one division to the next.

When the warp threads have been placed. insert the needle into the first division from the wrong side ; take up the first double thread, and put the needle back through the same division. Bring the needle up from the under side through the second division, take up the second double thread, and put the needle back through the second division. Continue this across the width of the model. Then put the needle through the first loop, but not through the cardboard, taking up one thread ; then take up the two threads, put the needle back into the loop from which the thread comes, and take up with this loop the one next to it together with a single thread, drawing them down to form a loop. Continue in this way until the card is filled, being careful not to fasten the darning at any point to the cardboard. This darning is used in knitted fabrics, and, like other darning, is always done over a piece of cardboard.

QUESTIONS AND ANSWERS.

What is the third model in this grade ? *Ans.* The knitted darning.

ow is the model prepared ? *Ans.* Cut a piece of cardboard three and one-half inches long and two and one-half inches wide ; draw lines one-half an inch from the top and bottom, and divide the model into eighth-inch spaces.

How is the warp thread put in ? *Ans.* With red cotton thread put a stitch in each division, first in the upper and then in the lower line, drawing the thread from one to the other, setting the needle in such a way that the thread is not straight, but slants from one division to the other.

How is the weft thread put in ? *Ans.* Insert the needle in the first division from the wrong side, take up the first double thread, and put the needle back through the same division. Bring the needle up from the under side through the second division, take up the second double thread, and put the needle back through the second division. Continue this the width of the model.

How is the weft thread brought back ? *Ans.* Put the needle through the first loop, but not through the cardboard, taking up one thread ; then take up two threads, put the needle back into the loop from which the thread comes, and take up with this loop the one next to it with a single thread, and draw down into a loop.

Is this way of putting the weft thread back and forth continued until the model is finished ? *Ans.* It is.

This darn is used for mending what sort of fabric? *Ans.* Knitted fabric.

How is a fabric of this kind prepared for mending? *Ans.* By basting a piece of cardboard under the place to be mended, and cutting the hole square.

In this case, is either the warp or the weft fastened to the cardboard? *Ans.* Neither.

QUESTIONS FOR REVIEW.

What is drafting?

What is a parallelogram?

What is the first thing to be done in drafting?

What is the gingham case when it is first drafted?

What is the second model of the third grade?

How many kinds of darning are there in this model?

What do the two threads in darning represent?

Are warp threads always straight?

What thread forms the pattern in darning and in weaving?

In what way is the linen darn different from the stocking darn?

How does the knitted darn differ from either of these?

How are these different kinds of darns used?

How is the fabric prepared that is to be mended with a darn?

Why should the place to be mended be cut square?

Why is cardboard basted under the fabric before the darning is done?

MATERIALS AND THEIR MANUFACTURE.

COTTON.

THERE was once a small black seed, which, with many others quite like it, was put into the ground one day in March on an island in the Atlantic Ocean, at the mouth of the Savannah River.

If this little black seed could have looked forth from its resting-place in the dark, moist ground, it would have seen a broad stretch of

water with low-lying islands all about, and close at hand the coast of
Georgia. But though it had not eyes, it had other wonderful natural
gifts, for it could draw different kinds of nourishment from the earth,
the air, and the sun; and these things enabled it to become something
so fine and so useful that it is not easy to believe the beginning was
only a little black seed.

First, it put slender fibers down deep into the ground in all directions. These little fibers had tiny mouths at their ends, which drank in
water and other food. After this a green shoot went straight up above
the top of the ground, and this put out small leaves and branches. All
the while the roots held the upper part firm, and gave it all that it
needed for nourishment. It was not long before green buds began to
show themselves, and soon pure yellow flowers, with reddish-purple
spots in the center, unfolded.

Little by little the seed became a shrub-like plant between three
and four feet high. When the pretty yellow flowers withered and fell,
green pods took their place. As time went on, these pods grew until
they were about as large as a small peach. When the pods turned from
green to brown they were ripe and burst open, and in each one was a
beautiful white ball of fine, soft fiber. This was the cotton.

The cotton family is a large one. No other kind of cotton is so
valuable as that which grows where the little black seed was planted.
It is called Sea Island cotton, and it is the very best in the world.
This is because the fiber, which is called the staple, is longer, finer, and
stronger than any other.

Another member of the cotton family is called New Orleans or
Upland cotton. Some of this has a green seed, and some a seed
that is gray-white. The blossom of this cotton is either pale yellow
or white; and the white fiber, or staple, about the seeds is shorter
than the Sea Island cotton. There are many other varieties. One,
which is called the Cuba Vine, has yellow fiber, out of which nankeen
is made.

To return to the story of the little black cotton seed. When the pod which held the fiber burst open, it was picked and taken to a machine called a cotton-gin. This occurred late in August, and you should have seen the field where this cotton plant grew. It was like a great silvery-white sea, and was one of the most beautiful sights in the world.

Besides the fiber in the cotton pod, there were a great many little black seeds—many more than it was necessary to keep for planting. These seeds must be taken out, and for this reason the fiber was all made to pass through the cotton gin. It was Eli Whitney of Massachusetts who invented the cotton gin. It can take the seeds out of three hundred pounds of cotton quicker than a man can pick them out of one pound.

Cotton Field.

The seeds were put into bags, taken to a mill, and made into a fine and useful oil. What was left after the oil was pressed out was put on the ground to enrich it, so

The Cotton Gin.

that what was planted in it would grow well. Part of the strong-fibered stalk of the cotton plant was used for making a basket, and the

rest for making a coarse sack. So every part of the plant was made useful, but the fine white fiber was by far the most valuable.

The cotton staple, which came from the plant of which the little black seed was the beginning, crossed the ocean, and went to Scotland, where it was made into thread, and then it came back to America and was sold.

CHAPTER V.

FOURTH GRADE WORK.

In this grade the work is altogether on garment fabrics; the knot is introduced, and the first garment of the course is drafted and cut. Up to this time the models have been small, and the necessity for a perfectly clean apron, used exclusively for sewing, has not been so great as it is now that the models are larger and more easily soiled. The apron, which is the third model of the grade, is a little work apron, which will be found a serviceable little garment, simple, and easy to cut and make, and one which can be utilized at every lesson.

If it be deemed desirable, the first model of this grade may be made the last of the preceding. This should not be done unless the work of the third grade has been so well done that more practice is superfluous, and there is still time for which no work is provided.

The pupil will be sufficiently advanced when the models of this grade are finished to work on silk, and so a simple but quaint little fancy bag of that material has been added to the models. This bag is not in the regular course of work, but is intended for those who complete the work of this grade in time to finish the bag before the close of the school year.

FIRST MODEL.

A length of gingham two inches wide and twenty-nine inches long, which should be divided into halves and quarters and marked, is first measured and cut by the pupil. Then thread a No. 8 sharp needle with No. 40 thread, doubled, and put in a line of running one-fourth of an inch from each edge. In doing this running, the stitches should be slipped off the needle without taking it out of the cloth, as by removing the needle a stitch is made uneven. Draw up the thread, twist it around a needle or pin placed at the end of the gathers; hold the right side of the work toward you, and place the gathers with a needle, holding the thumb directly over the gathers as they are placed. This must be done firmly and carefully. If any sound is heard, it shows that the needle is being drawn too roughly across the cloth and may injure it.

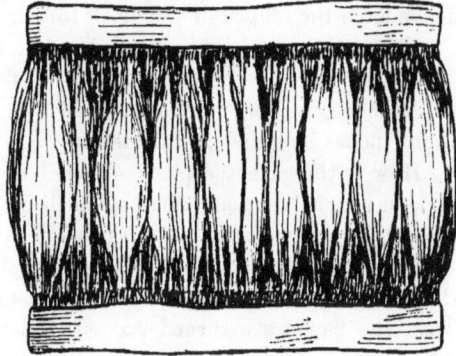

Then let the class cut two gingham bands eight inches long and two inches

Puff. First Model.

wide, and fit the gathered pieces to them. When the bands have been carefully basted onto the gathered piece, hem them on by taking up each gather as a stitch. Turn and hem down in the same way.

QUESTIONS AND ANSWERS.

What is the first model of this grade? *Ans.* A strip of gingham two inches wide and twenty-four inches long.

What is the first work on this model? *Ans.* To divide it into halves and quarters, and after marking it, to put in a line of running one-half an inch from each edge.

What number needle and thread is used? *Ans.* No. 8 needle and No. 40 thread.

How should the thread be prepared? *Ans.* It should be drawn between the thumb and finger of the left hand, or across a piece of wax, to prevent its kinking; and then doubled.

Is the needle taken out of the cloth in doing this running? *Ans.* It is not; the stitches are slipped off.

Why is the needle left in the cloth in running for gathering? *Ans.* If it is taken out it makes an uneven stitch.

Upon what does the length of the gathering stitch depend? *Ans.* It depends on the length of the cloth to be gathered.

When the running is finished, what is done? *Ans.* The threads are drawn up, and the gathers carefully stroked.

How long and how wide are the bands at the edge of this model? *Ans.* Eight inches long and two inches wide.

How is the gathered piece adjusted to these bands? *Ans.* The band is divided into halves and quarters, which are placed even with the divisions of the gathered piece after it is drawn up.

How are the three pieces put together? *Ans.* The bands are basted onto the gathered piece and then hemmed on, taking up a gather with each stitch. Then the bands are turned, basted, and hemmed in the same way on the other side.

SECOND MODEL.

The first of the two buttonhole models of this grade is a strip of felt cloth nine inches long and one inch wide, in which are cut eight button-holes an inch apart. Let the

Second Model. Buttonhole.

pupil practice cutting button-holes in a strip of drafting-paper until she can cut them properly. Then let the teacher cut the first one in the felt strip, and the pupil the others as she is ready to work them, until the eight are cut and worked.

The buttonholes should be worked with white No. 40 thread. As has already been said, the thread should be looped around the needle from right to left to form a pearl edge.

The second buttonhole model is a piece of white nainsook three inches wide and nine inches long. Double and overhand this along the side and the ends. Cut eight buttonholes one inch apart, overcast and work, as in the felt model, with No. 40 thread. Set four hooks and four eyes alternately between these buttonholes, sewing them on with a buttonhole stitch.

QUESTIONS AND ANSWERS.

What are the second models? *Ans.* Buttonhole models.

What is the first? *Ans.* A strip of felt nine inches long and one inch wide, in which eight buttonholes are cut.

How should these buttonholes be cut? *Ans.* Very straight.

How should they be worked? *Ans.* With No. 40 cotton thread which should be looped over the needle from right to left to form a pearl edge.

What is the second buttonhole model? *Ans.* A piece of nainsook three inches wide and nine inches long.

How is this prepared for buttonholes? *Ans.* It is folded together, overhanded at the sides and ends, and eight buttonholes are cut one inch apart.

How are these buttonholes worked? *Ans.* The same as those in the felt model.

How is the nainsook model finished? *Ans.* By putting on hooks and eyes between the buttonholes, arranging them alternately, and using the buttonhole stitch in sewing them on.

THIRD MODEL.

The third model is the gingham apron, which is the first garment drafted and cut. Two measures are taken, one across the chest from one arm to the other, and the other from the center of the chest to within an inch of the bottom of the skirt, which is the length. These measures should be taken by the pupil with the teacher's assistance.

When this is done, let the pupil draft a parallelogram twice as wide as the chest measure, and as long as the other measure. From the upper right-hand corner measure three inches down and two and one-half inches from the same corner to the left, and draw a curve for the arm scye. Cut two bands three and one-half inches wide, and as long as the chest measure. When the pattern has been drafted and cut, lay it on a double fold of the goods, pin and cut. Two of these pieces should be cut, one for the front and one for the back. The center of the back is cut open down the entire length.

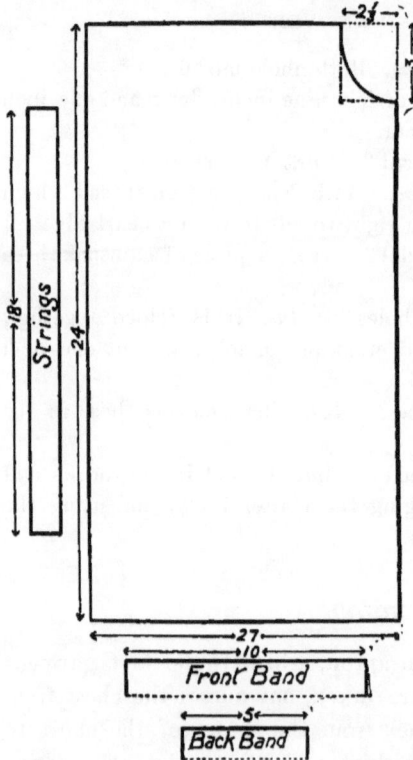

When the apron is cut, the under arm seams are basted and sewed in a very narrow seam, with three running stitches and one backstitch. The seams are then trimmed, turned, and back-stitched, making what is known as a French fell. Hem the two sides of the back in hems one-fourth of an inch wide, and the bottom in a hem an inch wide. Turn a hem a fourth of an inch wide about the arm scye.

Gather the top, beginning one and one-half inches from the arm scye. After stroking the gathers and basting on the bands, hem them to the apron by taking each gather up as a stitch, and hem them down in the same way. Turn a hem down the length of the string and across one end a quarter of an inch wide, and across the bottom one inch wide; sew on the shoulders, and tie.[1]

Model of Gingham Apron.

---

[1] It has been found that the white apron is more generally satisfactory for a sewing apron than the gingham one. The former may, therefore, be substituted for the latter if desired.

The work of this grade is finished by eight review lessons in practical darning without assistance from the teacher. First there should be the under-and-over stocking darn, not in canvas, but on a stocking. Let the pupils each bring a stocking that requires repairing. In the same way have linen and knitted darning practically applied. If there are pupils who have accomplished all the work of the grade in a satisfactory manner before the close of the year's work, let them make the silk bag as a reward of diligence.

THE SILK BAG.

Cut a straight strip of silk or ribbon fifteen inches long; if ribbon, six inches wide, if silk, seven inches wide. In each end of this piece of silk or ribbon, cut, two inches from the side edge, and three-quarters of an inch from the end edge, four buttonholes, lengthwise of the goods and three-eighths of an inch long. These buttonholes should be three inches apart.

Hem the two end edges and, if it is silk, the side edges. Overhand the two ends together an inch and one-half from each edge, leaving an open space of three inches in the center.

Silk Butterfly Bag.

Overhand the side edges together. Through the two buttonholes at each side run a narrow ribbon a yard long, so that it pulls up from both sides, and tie these ribbons in a bow on either side. As is obvious, two

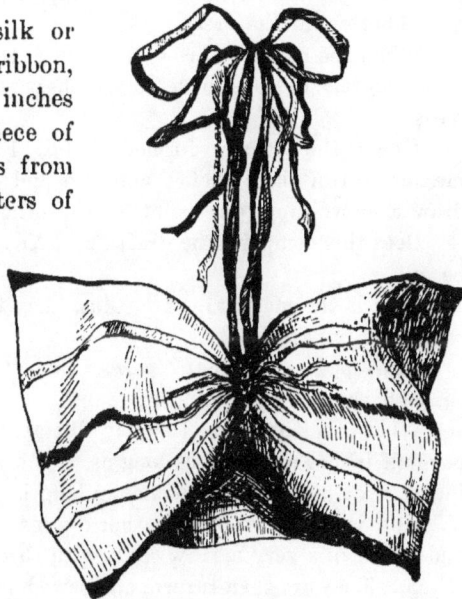

yards of ribbon are required. When it is drawn up, and the upper edges held together, it somewhat resembles a butterfly, and is often called by the children " The Butterfly Bag." It is a particularly convenient little receptacle to use in traveling, for buttons and other small trifles, as it can be laid perfectly flat, or hung up by the ribbon draw-strings.

QUESTIONS AND ANSWERS.

What is the third model of the fourth grade? *Ans.* The gingham apron.

How many measures are taken? *Ans.* Two, — one across the chest from one arm to the other, and one from the center of the chest to within an inch of the bottom of the dress.

When the measures are taken, how is the apron drafted? *Ans.* Draw a parallelogram twice as wide as the chest measure and as long as the other measure.

How is the arm scye formed? *Ans.* From the upper right-hand corner, measure down three inches and two and one-half inches to the left, and draw a curve from one point to the other.

Does this complete the drafting? Yes; and the paper pattern may now be cut.

How is the material cut? *Ans.* Lay the pattern on a doubled fold of the goods; pin and cut.

How is the back cut? *Ans.* Like the front, except that it is cut through the center the entire length.

What other parts are there to this apron? *Ans.* Two bands three and one-half inches wide and as long as the chest measure, and two strings three inches wide and eighteen inches in length.

How is the apron put together? *Ans.* The under-arm seams are basted, and sewed in a very narrow seam with three running stitches and one back-stitch. They are then turned, and sewed on the other side with a backstitch.

What is a seam finished in this way called? *Ans.* A French seam or fell.

How are the backs and bottom finished? *Ans.* The backs are finished with a quarter-inch hem, and the bottom with a hem an inch wide.

How is the neck finished? *Ans.* It is gathered across the front, except a space one and one-half inches from each arm scye. After the gathers are stroked, and the bands basted, hem them on, taking a gather to each stitch. Turn and hem them down in the same way.

How are the strings finished ? *Ans.* A hem a fourth of an inch wide is turned at the sides and one end, and one a half an inch wide at the other end.

Where are these strings placed ? *Ans.* They are sewed on at the end of the chest band, and tied on the shoulders.

How is the apron fastened at the top ? *Ans.* It is buttoned.

What is the last regular work of this grade ? *Ans.* Eight lessons in mending, with the different kinds of darning.

QUESTIONS FOR REVIEW.

Where is the knot first used ?
How is gathering done ?
How should buttonholes be cut ?
How should a buttonhole be overcast ?
How is the buttonhole stitch taken ?
What is the first garment drafted and cut ?
What is the shape of this apron before the arm scyes are cut ?
How many measures are taken ?
For what is this apron intended ?
How many different kinds of stitch are used in the apron ?
What sort of a seam is the under-arm seam ?
How is the kind of darn to be used determined ?

MATERIALS AND THEIR MANUFACTURE.

SPOOLS.

Up in the Highlands of Scotland, there grew a tall, slender, graceful tree, with shining white bark and delicate feathery green leaves. The name of this tree was birch; and it belonged to a very large family, which is found in all parts of the world where it is never very warm and often very cold. In Greenland there is no other kind of tree than the birch. This tree, which grew in Scotland, was one day cut down; and when the bark had been taken off, it was placed where it would get perfectly dry. It was then put on a wagon, and taken to a factory, where there was what is called a blocking machine.

A blocking machine is one which saws wood into blocks of any size that may be desired. When the wood of the birch tree had been cut, the blocks were put on a machine called a self-acting lathe, and in an instant they came out beautifully

Catkins and Leaves of Birch-Tree.

finished spools, all ready for the thread which later on was wound upon them.

Sometimes spools are made from the wood of ash-trees; but the largest number and the best are made of the birch, which is one of the most useful of trees. Its bark and leaves are used for medicine, and also for making yellow dye; the bark is made into drinking-cups, shoes, hats, and small boats called canoes; and there is also a fine oil made from it. The wood, because it gives forth a fresh, sweet fragrance when burnt, is used for smoking different kinds of meat and fish.

THREAD AND THE MANUFACTURE OF COTTON.

DID you ever think through how many hands the cotton must pass before it can become a nice, strong, smooth thread, several hundred yards long, and wound evenly upon a spool?

In the beginning, when the brown pods burst open, the cotton is as white as newly fallen snow; but by the time it has been picked, and has passed through the cotton-gin, and been prepared for shipment, it has gathered much grit and dirt, which must all be taken out.

After it has been thoroughly cleaned, it is placed upon feeding-tables, and from these tables it passes to big revolving rollers. These rollers are called cylinders, and are studded with strong teeth. As the cotton flies over them it looks like a great flock of white-winged birds; but it comes out from them in big sheets. It then passes over another series of cylinders, with small, sharp teeth, which make it into a fine white fleece just as thick in one place as in another.

As it comes from these cylinders, the pretty fleece is caught in a tube, and rounded into a coil so light and fragile that the least touch breaks it. And do you know, that if in any way the coil is broken, the machinery is so adjusted that it stops at once, and will not move again until it is perfectly joined.

When the cotton has been made into a little coil, it is put through the drawing-frame, where it is drawn out and doubled until all the fibers lie side by side. After this it is slightly twisted, and then wound on bobbins.

The thread I am telling you about was made in one of the largest factories in the world; and how many acres do you think it occupied? Between fifty and sixty acres. It is a pretty sight to see the white bobbins lying in long rows in the big building in which they are wound, ready for what is known as the spinning-mule. The spinning-mule is a machine mounted on a carriage that moves backwards and forwards; and the yarn is swiftly transferred from the fixed bobbins, and is twisted as it is wound onto the moving spindles. In this way several hundred

threads are twisted at the same time. Some of them are twisted for four and some for six cord thread. Some are fine and others coarse; but each number and kind of thread is made separately, although in the same factory.

When the thread is twisted as much as is necessary, and is ready to be finished, it is passed from the bobbins over a small peg of glass which acts as a guide, leading the swiftly flying thread to a little slit in an upright bar of steel called a cleaner. This cleaner detects a knot or unevenness of any kind, and at once stops the swiftly moving thread. In this way the thread is rendered absolutely without a flaw of any kind.

When the thread has passed through the cleaner, it is taken so rapidly through a flame of gas that it is not scorched, but all the little fibers on it are burned off. If you wish to see how it is done, take a piece of darning-cotton, and pass it very quickly through the flame of a lamp, and you will find that all the little fibers are burned away, but the cord is not injured.

When the thread has been passed through the flame, it is wound by a machine that fills a spool almost before you can see it. The spools are then labeled with a little round bit of paper on each end, on which is printed the kind of thread, where it is made, and the number. The spools are then packed in boxes, and are ready for the market.

Cotton which is to be used for making cloth is cleaned just as it is when it is to be made into thread. Before it is shipped to the factories, it is sent to great presses, where it is packed so tightly that it is almost as hard as a piece of wood. To clean it, and make it light and soft, the cotton is put six and sometimes seven times through a blower, inside of which is a beater of steel which turns many times in a second. Below it is a fan which revolves very fast, blowing out the dust, seeds, and sticks, and at last leaving it as light and white as sea foam.

There are two ways of spinning cotton, — one with the spinning-mule already described, and the other on the throttle or spinning-frame.

If you will take a piece of cotton cloth of the ordinary kind, such as calico or muslin, you will see that the threads go under and over each other just as they do in the first darn. The length-thread, which is the warp, is usually throttle spun; and the weft, which is the thread that goes from selvage to selvage, is spun on the mule frames. The warp threads are fixed on the looms; and in common cloth the weft is put under and over them by means of shuttles, which fly back and forth so rapidly that the eye can scarcely follow them.

The raising of cotton and its manufacture form two of the chief industries of the United States. Five-sevenths of all the cotton used in the world is raised in the Southern States; and a great deal of it is made into cloth and thread in the great factories of this country.

CHAPTER VI.

WORK OF THE FIFTH GRADE.

THERE are four models in this grade, two of which are garments. The latter should be drafted, cut, and finished without assistance from the teacher, who simply directs what is to be done, as the pupil should now be capable of working from clearly defined, explicit directions. In cutting garments, let it be borne in mind that the right-hand side is in every instance the back part of the garment.

THE FIRST MODEL.

The first model of this grade is a square of rather coarse linen, measuring six inches on each side, which is for practice in finer hemstitching than has been done before, and the first work in linen marking.

Three-quarters of an inch from the edge on each of the four sides of this square of linen draw three threads. Along the length of the model, after creasing down an eighth of an inch, turn a hem that just reaches the line of drawn threads. When these hems are basted, turn and baste the width hems in the same way. The two lengths, and after that the two width sides, are turned so that the four corners may be uniform, each one being square.

Let this edge be hemstitched by taking up three threads at each stitch. This being done, have each pupil write or print her name in pencil in the left-hand lower corner as a guide for the marking; stitch this name in white marking-cotton. Then go over it in overhand backstitch in red marking-cotton, picking up each white backstitch, and the model is finished.

QUESTIONS AND ANSWERS.

What is the first model of the fifth grade? *Ans.* A piece of linen six inches square to be hemstitched and marked.

How is it prepared for hemstitching? *Ans.* Three threads are drawn three-quarters of an inch from each edge of the model.

How is the hem turned? *Ans.* It is first turned in an eighth of an inch along the length edges, and the hem is turned to meet the line of drawn threads. The width edges are then turned in the same way, so that a square is formed at each corner.

Why are the two length edges turned and then the width edges? *Ans.* That the corners may be uniform.

When the hemstitching is finished, how is the marking done? *Ans.* The name is first written or printed with a pencil.

What is the next step? *Ans.* The name is then carefully outlined in backstitching with white marking-cotton.

How is it finished? *Ans.* Overhand the white backstitching with red marking-cotton, taking up each stitch of the backstitching.

SECOND MODEL.

The second model in this grade is a little fancy sewing apron, which may be made of white cambric, barred mull, nainsook, or any kind of light printed goods. There are three measures taken, — a waist measure, a measure from the waist line on the left side over the shoulder and across the back to the waist line on the right side, and the length of the dress skirt less two inches.

For the skirt of the apron, take the length of the dress skirt less two inches, and cut it thirty inches wide. For the band at the waist, cut a strip lengthwise of the goods, one inch longer than the waist measure, and four and one-half inches wide. The one inch is added to the waist measure to allow for the lap. Cut the shoulder pieces also lengthwise of the goods, two and one-half inches wide, and as long as the measure.

Cut the ruffles one inch wide for the shoulder pieces, allowing half as much more than the length for the fullness. After gathering these ruffles, and laying the gathers, baste them on either side of the shoulder pieces. Hem the ruffles on to the shoulder pieces; turn and face with a three-quarter inch bias facing. Explain that bias facing is preferable, as a rule, to straight, because it is more elastic.

Turn up the lower edge of the apron skirt in an inch-wide hem. Finish the back edges with a half-inch hem. Gather this skirt, and baste on to the waistband so that the two back edges are three inches from the ends of the band.

Sew the shoulder pieces onto the waistband with the two edges coming close together in the center of the front. Work a buttonhole in the other two ends of the shoulder pieces, cross in the back, and button an inch and a half each side of the center of the back onto the band. Finish the band with a button and buttonhole.

PLAIN SEWING APRON.

The lower part of this sewing apron is like the one already described. For the upper part, take a measure from the highest part of the shoulder to the waist, and another from one arm to the other across the fullest part of the chest. The first measure is used for length, and the other for width. Fold the goods to be used lengthwise, and cut an oblong piece according to the measures. Curve the upper part for the neck. Turn a quarter-inch hem along the sides, and face the neck curve with a bias piece of the goods. Gather the lower part one-quarter of an inch from the edge, and stroke carefully. Draw the gathering-thread until it measures five inches, and fasten. Place the center of the gathers on the center of the apron band, and baste. Stitch, add a piece of finishing-braid, and hem down on both sides. Make a buttonhole in the right side of the apron band, and place a button on the left side. The upper part of this little apron is held in place by safety pins at the shoulder.

What is the third model of the fourth grade ? *Ans.* A sewing apron.

How many measures are taken for this apron ? *Ans.* Three — the waist measure, the length from the waist over the shoulder to the waist on the other side, and the length of the skirt of the dress less two inches.

How is the skirt of this apron cut ? *Ans.* As long as the skirt measure, and thirty inches wide.

How is the waistband cut ? *Ans.* Four and one-half inches wide, and one inch longer than the waist measure.

Why one inch longer than the waist measure ? *Ans.* To allow for the lap.

How are the shoulder pieces cut ? *Ans.* As long as the measure, and two and one-half inches wide.

How are they finished ? *Ans.* With an inch-wide ruffle on each side, and around one end.

How much longer than the bands are these ruffles cut ? *Ans.* One-third longer than the bands, with four inches added for the two ends, in which the buttonholes are placed.

How is the skirt finished at the bottom and sides ? *Ans.* With a two-inch hem at the bottom, and a half-inch hem at the sides.

How is the skirt put onto the band ? *Ans.* First fold the two ends of the band together and mark. Then gather the skirt, and when the gathers have been stroked, baste the skirt so that each side is three inches from the center of the band at the back.

How are the shoulder pieces placed ? *Ans.* Put the two ends on in such a way that the edges meet in front, and hem them on the under side of the band. Cross in the back so that the piece from the left side of the front is buttoned on one and one-half inches from the right side of the back, and the other in the same way, one and one-half inches from the left side of the back.

How is the apron finished ? *Ans.* With a button on the right side and a buttonhole on the left side of the belt.

THIRD MODEL.

More difficult than anything that has yet been cut and drafted are the drawers of this model. As in the apron, the first thing is to draft and cut a paper pattern. First take two measures, — a loose waist measure, and the length from the waist to the knee.

Draw a parallelogram with half the waist measure, to which four inches have been added for the two horizontal lines, and the distance from the waist to the knee for the vertical lines. The four inches added to half the waist measure are for gathers. As the four lines of the parallelogram are help lines, draw them dotted. Let the base line of this parallelogram be A, the left-hand vertical line B, the upper horizontal line C, and the right-hand vertical line D. In these drawers, as in all the garments of this system, the right-hand side is the back.

Draw a dotted vertical line, E, through the center of the parallelogram. Draw a dotted horizontal line, F, one inch below half the distance between A and C, extending it four inches on each side beyond lines B and D. Draw a dotted oblique line from the end of line F to A on either side, and within these a slightly curved cutting-line. Two inches below line A draw a straight drafting-line three-fourths of an inch longer than line A on each side, and connect with line A by an oblique line. This forms the hem at the bottom.

From the end of line F, on the right-hand side, which is to be the back of the drawers, draw a straight dotted line, H, that extends two and one-half inches above line C. Draw a dotted line, I, four inches to the left, and connect line C. From the center of line I, draw an oblique cutting-line to the end of line F, and another connecting it with dotted line E, which divides the back from the front.

For the front, extend line C one inch, and from this point to the end of line F draw an oblique line. From the end of line C to the vertical line E make the dotted line C a cutting-line. From line C, five inches down line E, cut an opening for a placket, if the drawers are to be closed; if they are to be open in the center, this is not necessary. When this pattern has been drafted and cut by the pupil until she can do it with perfect ease, let the pattern be laid on the doubled goods, and the drawers cut. The material used for these drawers should be Lonsdale muslin of good quality.

Take a piece of Lonsdale muslin six inches square, fold diagonally and cut, slightly curving the seam. Baste the seam one-half an inch

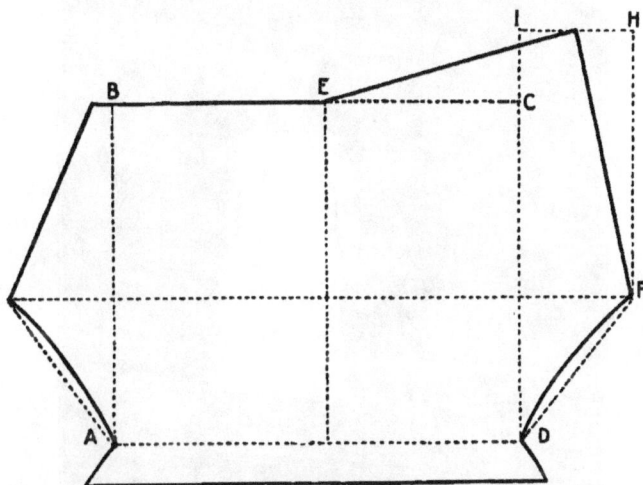

Model of Drawers.

from the edge, backstitch, trim one edge so that it can be turned under, baste and fell. This is to teach the child how to sew and finish a bias seam.

The pupils are now ready to put the drawers together. First the curved seam that forms the lower part of the leg is basted a half-inch from the edge, is backstitched, trimmed and felled as the seam just finished was. In basting these seams, care should be taken to have the

Child's Drawers and Underwaist.

two fronts face each other, and the reason for this should be explained to the pupils.

Turn and fell the hem at the bottom of the leg. Then join the center seam so that the two leg seams are exactly opposite each other. If desired, this seam may be joined three inches from the top in front, and from that point faced and left open.

To finish the placket, take a piece of muslin ten inches long and one and one-half inches wide. Seam this around the placket so that the seam is on the right side. Turn in the edges of this strip, and bring over and fell. The fullness where this strip is turned forms a sort of gusset.

For the front band, add one and one-half inches to half the waist measure for the length, and cut it four and one-half inches wide. The back band is the same width, but is one and one-half inches shorter than half the waist measure. Both bands should be cut lengthwise of the goods.

The gathering of the front and back should begin four inches from each side. When the gathering is done and the gathers are stroked, the bands should be basted on and felled, taking up a gather with each stitch; then turned, basted, and felled in the same way. There should be a buttonhole in the two ends and the center of each band, making six in all. These drawers are to be buttoned onto an underwaist.

QUESTIONS AND ANSWERS.

What is the third model of the fifth grade ? *Ans.* A pair of drawers.

What is the first thing to be done? *Ans.* Draft and cut a pattern.

What is the first thing to be done in preparing to draft a pattern? *Ans.* Take the measures.

How many measures are taken for drawers ? *Ans.* Two — a loose waist measure, and the length from the waist to the knee.

How is the pattern drafted from these measures ? *Ans.* First a parallelogram is drawn with half the waist measure, to which four inches is added for gathers, for the base horizontal lines, and the distance from the waist to the knee for the vertical lines.

Is this whole parallelogram in cutting-lines? *Ans.* No; only the base line A and the right-hand vertical line D are cutting-lines, the lines B and C being help lines.

What is the next line? *Ans.* A dotted vertical line, E, through the center of the parallelogram.

What is the next step? *Ans.* Draw a dotted horizontal help line, F, one inch below half the distance between A and C, extend it four inches on each side beyond lines B and D.

How is the drafting continued? *Ans.* Draw a dotted oblique help line from the end of line F to A on either side, and within these a slightly curved cutting-line.

How is the hem provided for? *Ans.* Two inches below line A draw a straight drafting-line, three-fourths of an inch longer than the line on either side, and connect the two with an oblique line.

Which side of the pattern is the back? *Ans.* The right-hand side.

How is the back drafted? *Ans.* From the end of line F on the right-hand side, draw a vertical dotted help line H, extending two and one-half inches above line C, and from the end of this line draw a dotted help line, I, four inches to the left, and connect with line C.

How are the next lines drawn that finish the back? *Ans.* From the center of line I, draw an oblique cutting-line to the end of line F at the right, and another to the end of line E at the left.

How is the front drafted? *Ans.* Extend line C an inch, and from this point draw an oblique line to line F.

What finishes the drafting of this pattern? *Ans.* Make the help line C from E to B a cutting-line, and cut down line E five inches for a placket.

What sewing is done before the drawers are put together? *Ans.* A diagonal seam through a six-inch square of muslin is cut, sewed, and felled.

What is this for? *Ans.* To show how a bias seam is sewed and felled.

How are the drawers put together? *Ans.* The curved seams that form the lower part of the leg are basted a half-inch from the edge, care being taken to have the two fronts come together.

Why should the two fronts come together ? *Ans.* Because otherwise both parts will be for one leg.

How are these seams sewed ? *Ans.* They are first backstitched and then felled.

What is the next thing to be done ? *Ans.* Turn the hems at the bottom.

How is the center seam finished ? *Ans.* It is either basted, with the leg seams exactly opposite each other, backstitched and felled, or it is joined three inches in the front, and then faced and left open.

When the center seam is closed, where are the plackets placed ? *Ans.* On each side ; and they are five inches long.

How are the plackets finished? *Ans.* Take a piece of muslin ten inches long and one and one-half inches wide, and seam this on around the plackets so that the seam is on the right side. Turn in the edge of this facing, bring it over on the seam, baste, and fell.

What does this fullness where this facing is turned form? *Ans.* A kind of gusset.

How are the bands cut ? *Ans.* For the front band, take half the waist measure, and add one and one-half inches for the length, and make it four and one-half inches wide. The back band should be the same width, but one and one-half inches shorter than half the waist measure.

If the drawers are not closed, how should the band be cut ? *Ans.* Four and one-half inches wide, and as long as the waist measure, with one inch added for the lap.

How should the bands be placed ? *Ans.* The gathering should begin four inches from each side, and when the gathers have been stroked, fell on the bands, taking up a gather at each stitch, then turn, and fell in the same way.

How are the buttonholes placed ? *Ans.* If the drawers are closed, and two bands are used, there is one buttonhole in the end of each band and one in the center, making six in all.

If one band is used, how are the buttonholes placed ? *Ans.* There is one in the front, one at each side, and two behind, making five in all.

FOURTH MODEL.

.The fourth and last model of this grade is a five-inch square of flannel in which is cut a right angle opening, the shape of a tear.

Fourth Model.   Right-angled Tear.

First baste this piece of flannel onto a piece of cardboard, then, with silk the exact color of the material, darn straight across in a very fine running stitch, taking up the nap only, carefully drawing the thread just even with the cloth which is being mended. Continue this until the incision is perfectly closed.

When this model is completed, let the pupil review all the darning that has been taught, not on canvas, but on fabrics. The work should be prepared and completed without assistance from the teacher, and with the neatest precision, before the pupil is permitted to begin the work of the next grade.

QUESTIONS AND ANSWERS.

What is the last work in this grade ? *Ans.* A flannel darn and a review of all the darning.

What is the flannel or tear darn ? *Ans.* It is a three-cornered opening representing a tear.

How is it prepared for darning ? *Ans.* It is basted very carefully onto cardboard, in such a way that it lies perfectly smooth, with the edges together.

How is the darning done ? *Ans.* With silk the exact shade of the goods, and with stitches that take up the nap only, and are very close together.

Do the stitches run both ways? *Ans.* No; they only extend straight across the opening.

How many kinds of darning have now been taught? *Ans.* Four kinds, — stocking darning, linen darning, the knitted darn, and the tear darn.

QUESTIONS FOR REVIEW.

What is the first work of the fifth grade?

Of what use is hemstitching?

For what is the marking taught in this grade used?

Explain how linen hemstitching is done.

How is marking done?

What is the second model of the fifth grade?

How is it trimmed?

In cutting ruffles, how much fullness should be allowed?

Should facings, as a rule, be bias or straight?

How many measures are taken before beginning to draft drawers?

What are they?

What geometrical figure is first drawn in drafting a pattern for drawers?

Draft a pattern for drawers from measures given.

What are help lines, and how are they drawn?

What are cutting lines, and how are they drawn?

Is the pattern laid on a single or double piece of the goods when cutting the drawers?

What is a fell?

In sewing up drawers, how can one be sure that the two parts will not be both for one side?

What is a placket?

How is a placket faced and finished?

What darn is taught in this grade?

In what way is this darn different from the stocking and the linen darns?

How many darns have been taught ?

What are they ?

Why is it necessary to know different sorts of darns ?

How must all darns be prepared ?

MATERIALS AND THEIR MANUFACTURE.

HOW SILK IS MADE.

SILK is not only a very beautiful, but a very wonderful fiber, for it is made either by insects or worms. There are many insects that make themselves little houses out of silk spun from their bodies. The webs and nests of spiders are of silk.

The silk fiber from which fabrics are made is spun by the mulberry silkworm, and beautiful cloth was first made from it by the Chinese, farther back than we have record of. They did not want anybody to know their art; and they kept it such a secret that every one supposed that the cloth was made from some kind of a plant, like flax or cotton. At last a traveler, about the year 550 A.D., found out the secret, and brought away some eggs of the silkworm in a hollow bamboo cane. These eggs were hatched, and in this way silk culture became known to all the world.

The famous writer and Greek philosopher, Aristotle, in speaking of the silkworm, says that it is "a great worm that has horns, and so differs from other worms." [1] This big worm, when it is full grown, first spins a web about itself of finest fiber, often four thousand yards in length. The worm moves as it spins, in such a way that the fiber is wound round and round as regularly as thread is wound onto a spool. In three days the house of silk is complete. Then the worm lies still until it becomes a moth, which is similar to a butterfly. This moth moistens the silk house, which is called a cocoon, and makes its way

[1] Ask the children if they think that Aristotle was correct in saying that other worms have not horns.

out.. Very soon after the moth leaves the cocoon it begins to lay eggs, and in three or four days has laid from four hundred to seven hundred.

The eggs of the silk-moth are carefully put into trays, and kept where the temperature does not vary, being neither too warm nor too cold; and soon the little worms begin to hatch. A paper punctured full of small holes is laid over the trays, in order that the worms may crawl through these holes. In this way fragments of shell, which adhere to them and would kill them, are scraped off.

As soon as the silkworms are freed from their shells, they begin to eat, and they do nothing else all day. Their food is mulberry leaves, and the worms hatched from an ounce of eggs will eat a ton of leaves in a month. The worms change their coats almost every week. At the end of a month they are full grown. They then creep up on branches pro-

Silkworm, Cocoon, and Moth.

vided for them, and begin to spin silk houses for themselves, in which they become moths. Only a few of these moths are permitted to live, and eat their way out of the cocoon, for that injures the silk. Enough to lay eggs are left on the branches; the others are removed and killed by baking the cocoons in an oven, exposing them to the hot rays of the sun, or shutting them up in a close room where charcoal is burning.

It is a great deal of work to care for silkworms, and where labor is valuable they are not very profitable. Then, too, they require so much to eat that they can only be successfully cultivated where there are great plantations of mulberry-trees.

THE MANUFACTURE OF SILK.

The silk which the worm has spun is as fine as the web of a spider before it is unwound. The cocoons are assorted, and those of similar color are placed together. When this has been done, they are put into tepid water. If the water is too cold, the gum of the cocoon will not soften enough to permit the fiber to unwind well; and if it is too warm, it will sink to the bottom. Girls who are experts stir the cocoons until they soften, and the end of the fiber is found.

Silk Winding.

A number of these delicate silk fibers are put together through an eyelet, and after being crossed and twisted are wound on a reel. When these threads are dried, they cling together, and form a compact fiber of raw silk.

From the reels, this silk fiber is wound upon bobbins in such a way that the threads are all in diagonal lines. These bobbins are next placed on the spinning-frame and slightly twisted. Then these strands are cleansed, wound together upon a reel, and twisted into one thread.

The thread is then reeled into big skeins; and as it is moist, it must be thoroughly dried, thus making it ready to be sold to manufacturers by the pound. As these skeins are somewhat stiff, they are whirled about in hot soap and water to make them flexible. They are then dried, packed into linen bags, boiled in water, and again dried. The silk is now white and soft, and is ready to be sent away to be

Reels and Skeins of Silk.

colored, and woven into ribbons or some of the many different kinds of beautiful silk cloth.

Spun silk is made of the waste silk and poor cocoons. It is not reeled, but is separated by machinery into strands about a foot long. These are spun together as cotton is, and made into yarn called spun silk, which is by no means as durable as the other kinds of silk.

There are over two hundred silk mills in this country; but most of the silk used here is brought from China, as there are not many silkworms raised in the United States.

CHAPTER VII.

SIXTH GRADE WORK.

THE mechanical process of cutting garments by chart, which has been so long in use, seems, upon first examination, to be much more simple and teachable than the scientific method of this system. That it is not so has been abundantly proved by repeated and continued experiments with hundreds of children. Since the understanding of general laws makes all things plain, when the principal facts upon which this system is based are understood, the process is found to be as simple as it is reasonable. The system does indeed tax the understanding at every step, and it is the aim of its authors that it should do so. Anything less than this would defeat its chief object, which, as has been repeatedly affirmed, is to incite independent constructive thought.

While the aim of the system is primarily educational, it has also, as a means to an immediate end, advantages which are easily demonstrated. The subtle philosopher Amiel, in his famous journal, declares that every human being is a unique example, and should be so considered, and that satisfactory results cannot be obtained in any other way. This certainly is true in fitting the human form. It is of course possible to strike a general average; but when it comes to that nicety which distinguishes excellence, it can only be obtained by considering each individual as separate and exceptional. The scientific system of garment cutting makes this consideration of the individual one of its fundamental principles, as will be seen in the drafting of the waist of this grade. As a result, the work of fitting is almost entirely eliminated. · In the school where this system has been successfully taught for the past six years, in more than one instance the graduating gown, which finishes the course, has been completed without being fitted, and proved in every way perfectly satisfactory.

Although the waist of this grade is a simple underwaist, yet as it is the foundation of all others, the drafting and cutting of it should be very thoroughly understood. The measures should be taken and the pattern drafted until it can be done with the utmost ease, without suggestion from the teacher.

CHILD'S UNDERWAIST.

First there are eight measures to be taken as follows ; A bust and waist measure ; a front measure, which is taken from the hollow of the neck to the waist; a front width, which is one-fourth of the bust measure ; a back length, which is taken from the neck to the waist; a back width, from one arm to the other ; a side length, from under the arm to the waist; and a shoulder measure, from the neck to the point of the shoulder. If, as occasionally happens, the pupil is not equal to the mathematical calculations where it is necessary to divide parts of inches, each fraction of an inch in the measures may, for convenience, be made a whole one. This must be done by adding the part of an inch required to make the measures in even inches. This is not recommended, and should not be done when it can be avoided, as garments cut from patterns drafted in this way require much more fitting than when the exact measures are used.

When the measures have been taken and tabulated, draw a parallelogram, with half the bust measure for the base line A, and the front length with two and one-half inches added for the vertical line B. Draw the second horizontal dotted help line C and the dotted vertical line D, which completes the parallelogram. The help lines are drawn dotted to distinguish them from the cutting-lines which later on form the outline of the pattern.

Measure the side length from line A on the vertical lines B and D, and from these two points draw a dotted help line E. Measure the front length from line A on the vertical lines B and D, and draw a dotted help line F. Measure one-fourth of the bust measure on line A from the left-hand lower right angle of the parallelogram, and also on line C from the corresponding upper right angle, and draw a straight dotted help line G.

Child's Underwaist.

For the back, which is always the right-hand side, take half the back width, and measure it on the base line A from the lower left-hand right angle, and on C from the upper left-hand right angle of the parallelogram, and between these points draw a dotted help line H. From the angle of lines C and D down line D, measure one-half inch, point 1, and along C an inch and a half to point 2, and connect with a slightly curved cutting-line. From point 2, draw an oblique cutting-line to the intersection of lines H and F. Measure off the length of the shoulder on this line to point 3. From this point, draw a slightly curved line to the intersection of lines E and H, which forms the back arm scye. Add three-fourths of an inch to lines A and C at the right, and draw a dotted vertical help line I. Again add an inch to each of these lines, and draw a vertical cutting-line J. These additions of an inch and three-fourths to each side of the back are for the lap and the buttons; when the waist is finished, the two edges of the back should come exactly together at line D.

Measure two and one-half inches down line B from the angle of lines B and C to point 4, and the same distance on line C to point 5, and connect with a curved line which forms the neck. From point 5, draw an oblique line to the intersection of lines F and H, and, from the neck, measure the shoulder length to point 6. From point 6 to the intersection of lines E and H, draw a curve for the front arm scye.

A child's waist measure is often larger than the bust; when this is the case, add whatever the waist is in excess of the bust on both sides of line H below the arm scye. If the bust measure is more than the waist measure, lay off one-half of the difference on either side of line H along line A, slanting in the shape of a dart along line H from the intersection of lines E and H.

As has already been said, this pattern should be drafted and cut by the pupil until it can be done with perfect facility. Then let the pattern of the front and back be laid on a double fold of the goods, and cut, allowing one-half an inch on the sides and shoulders for seams. Baste the shoulder and side seams along the line of the tracing, and after backstitching, trim one seam and fell. Face the neck and arm scyes with a bias piece one inch wide. Finish the bottom with a deep facing, and set buttons to correspond with the buttonholes in the bands of the drawers. To cut bias pieces for facing this waist or for any other purpose, begin at the corner, and fold back two inches. Fold over and over until a piece as long as the strips required is folded.

Divide the strip thus folded into parts as wide as is desired, and cut through. In this way long strips may be cut at one stroke of the shears.

The back is finished by turning down the right side one and one-half inches, and the left side one inch. The buttons on the left side should be set three-fourths of an inch from the edge, and the buttonholes one-fourth of an inch from the right side. This brings the waist together along line D.

UNDERWAIST OF MANILLA PAPER.

In order that the pupil may gain practice without waste of material, let the measures be taken and reduced to quarter inches, and a waist be drafted and cut of manilla paper. In drafting this paper waist, allow one-eighth of an inch for the side and the shoulder seams. When it has been drafted and cut, join the seams by backstitching. After making a few of these, the pupil will be able to handle materials with ease, and the waste which mistakes entail will be avoided.

If the difference between the bust and waist is four inches, one dart is sufficient, and there is no slant at the back; if the difference is five inches, there should be one dart, and one inch slant at the back; if the difference is seven inches, there should be two darts, and one inch slant at the back.

QUESTIONS AND ANSWERS.

What is the first work of the sixth grade? *Ans.* Drafting, cutting, and making an underwaist.

How many measures are taken for an underwaist? *Ans.* There are eight measures taken for the underwaist.

What are they, and in what order are they taken? *Ans.* They are taken in the following order: The bust and waist measure; the front length from the neck to the waist; one-fourth of bust measure; the back length from the neck to the waist; the back width from one arm to the other; the side length under the arm; the length from the neck to the point of the shoulder.

Can this pattern be cut without using parts or fractions of inches? *Ans.*

Yes; by making the measure a whole instead of a part of an inch. The part of an inch must always be added and not subtracted.

Will a waist fit as well when the measures are made in even inches? *Ans.* No; it will require much more fitting.

What is done with the measures as they are taken? *Ans.* They are set down at the right on the drafting-paper in the order in which they are taken.

After the measures are taken, what is the next thing to be done? *Ans.* Draw a parallelogram with half the bust measure for the base line A, and the front length with two and one-half inches added for the vertical line B.

What sort of lines are C and D of this parallelogram? *Ans.* They are dotted help lines.

Why are help lines drawn dotted? *Ans.* To distinguish them from the cutting-lines that form the outline of the pattern.

What is the next thing to be done? *Ans.* Measure on lines B and D from line A the side length, and draw a dotted help line E.

Which is the next measure used? *Ans.* The front length is measured from line A on lines B and D, and the dotted help line F is drawn from one to the other.

What is the next step in drafting this waist pattern? *Ans.* Measure one-fourth of the bust measure on line A from the left-hand lower right angle and from the left-hand upper right angle on C, and draw a dotted help line from one to the other.

Which side in this system is always the back? *Ans.* The right-hand side.

How is this pattern for the back drafted? *Ans.* Take half the back width on lines A and C, measuring from the upper and lower right-hand angles, and from these points draw the dotted help line H, and from the angle of lines C and D measure one-half an inch down line D, point 1, and an inch and one-half along line C, point 2, and connect with a slightly curving line for the back of the neck.

How is the shoulder line obtained? *Ans.* From point 2 to the intersection of lines H and F draw an oblique line, and mark the shoulder measure, point 3.

How is the back arm scye obtained? *Ans.* A slightly curved line is drawn from point 3 to the intersection of lines E and H.

How is the lap for the buttons and buttonholes at the back formed? *Ans.* Add three-fourths of an inch to lines A and C at the right, and draw a vertical help line I; again extend lines A and C one inch, and draw a vertical cutting-line J.

How is the right side of the back of the waist finished? *Ans.* By turning in an inch and one-half and hemming.

· How is the left side of the back finished? *Ans.* By turning in one inch and hemming.

Where are the buttonholes cut? *Ans.* In the left side, one-quarter of an inch from the edge.

Where are the buttons set? On the right side, three-fourths of an inch from the edge.

How is the curve for the front part of the neck drafted? *Ans.* Measure two and one-half inches down line B from the angle of C and B, point 4, and the same distance on line C to point 5, and connect with a curved line.

How is the shoulder of the fronts drafted? *Ans.* From point 5 draw an oblique line to the intersection of lines F and H, and from the neck lay off the shoulder measure to point 6.

How is the front arm scye obtained? *Ans.* From point 6 to the intersections of lines E and H, draw a curve.

If the waist measure is larger than the bust measure, what change is made in the pattern? *Ans.* Half the difference is added along line H, gradually slanting to the arm scye.

If the bust measure is more than the waist how is the pattern changed? *Ans.* One-half the difference between the waist and bust measures is laid off on line A from line H, these points being connected by slanting lines with the point of intersection of lines E and H.

When the pattern has been drafted and cut, how should the waist be cut? *Ans.* It should be laid on a double width of the goods and cut.

How much should be allowed on the shoulder and side seams? *Ans.* One-half inch on each.

How should this waist be put together? *Ans.* The side and the shoulder seams should be basted, backstitched, trimmed, and felled.

How should the neck and arm scyes be finished? *Ans.* With bias facings, an inch wide.

How should these facings and all bias pieces be cut? *Ans.* Begin at the corner of a piece of cloth and fold back two inches, then fold over and over until a piece as long as the strip required is folded. Divide this strip into parts as wide as is required, and cut through.

How is the bottom of the waist finished? *Ans.* With a straight facing, two inches wide.

How is the back of this waist closed? *Ans.* With button and buttonholes.

How are the buttons on the lower part of the waist set? *Ans.* To correspond with the buttons in the band of the drawers.

UNDERSKIRT WITH SHOULDER STRAPS OR WAIST.

For this underskirt take the length one inch shorter than the dress-skirt, and to this add two inches for a hem. For a child of from three to five years of age, take two and one-half widths of cambric or Lonsdale muslin, and for a child of from five to twelve years, take three widths of either of these materials.

Join the seams of the skirt in what is known as a French fell, by first sewing up with three running stitches and one backstitch; then trim, and turn, backstitching the seam on the other side.

Take the center of a width for the front, and, directly opposite, cut a placket five inches deep. Turn a half-inch hem on the right side, and a quarter-inch on the other; lap the right side over the left, and stitch at the bottom.

For the band, cut a strip of goods six and one-half inches wide, and one-half an inch longer than the waist measure. This, when it is seamed on and doubled, makes a band three inches wide. Divide this band and also the skirt into four equal parts; gather the two fronts on one thread, and each of the back divisions on a separate thread. Lay the gathers, and baste onto the band, bringing the center of the front of the band and skirt together, and the side divisions of the skirt one-half an inch back of the side divisions of the band.   Fell on, taking up a gather with each stitch, then turn and hem down in the same way.   Close with buttons and buttonholes.

Cut the shoulder straps three inches wide, double, and overhand them together.   The two ends of these straps are felled onto the band together, and either side of the front on an inch slant.   These straps are brought over the shoulders, and fastened to the band one inch from the center of the back on either side.

When this skirt is made with the waist, the latter is not faced, but is divided like the band, and is turned in and felled onto the skirt, and then faced with a two-inch facing.

The work of this grade is finished with eight lessons in practical darning.   This work should be done without any assistance from the teacher, the pupil selecting the sort of darn suitable for the fabric to be mended, and also the wool or thread with which the work is to be done.   Let nothing less than perfect work, independently done, be accepted.

QUESTIONS AND ANSWERS.

How long should an underskirt be cut ?   *Ans.* One inch shorter than the dress skirt, with two inches added for a hem.

How wide should it be ?   *Ans.* For a child from three to five, two and one-half widths of Lonsdale muslin or cambric should be used, and for a girl from five to twelve years, three widths of these materials are required.

How should this skirt be seamed up ?   *Ans.* It should first be sewed up on the right side with three running stitches and one backstitch, and then trimmed, turned, and sewed onto the other, making a French seam.

Where is the placket placed, and how is it finished ?   *Ans.* Take the middle of a width for the front, and directly opposite cut a placket five inches deep, turn a half-inch hem on the right side, and a quarter-inch on the other, lap the right over the left, and stitch at the bottom.

How is the band cut ?   *Ans.* Take a strip lengthwise of the goods, six and one-half inches wide, and one-half an inch longer than the waist measure, and divide it into four parts.

How is the skirt gathered ? *Ans.* First divide into four parts and notch. Gather the front half on one thread, and the two back parts, each on a separate thread.

How is the skirt put onto the band ? *Ans.* After the gathers are stroked, put the front notch in the skirt even with the corresponding notch in the band, and the side notches in the skirt one-half an inch back of those in the band, baste and fell, taking up one gather with each stitch.

How is the buttonhole placed ? *Ans.* On the right side of the band near the end, so that it laps from right to left.

How are the straps placed ? *Ans.* They are hemmed onto the inside of the band in front, with the edges together, so that they can be crossed on the chest, and are brought over the shoulders, and felled onto the band on each side one inch from the center.

How are the shoulder bands cut and made ? *Ans.* They are lengths of cloth three inches wide, doubled, and overhanded together.

When the skirt is a part of the underwaist, how is it finished ? *Ans.* The waist is not faced, and the skirt is put on in the same way as on a band, and the facing is then felled onto the skirt and waist.

What finishes the work of the sixth grade ? *Ans.* Eight practice lessons in the four kinds of darning that have been taught.

If time permits, some of the garments which the pupil has learned to make may be made for the doll of the department before the work of this grade is closed.

QUESTIONS FOR REVIEW.

What is the first thing to do in drafting a waist pattern ?

How many measures are taken ?

What are they, and in what order are they taken ?

When the measures are taken, what is the first thing to be done in drafting a waist ?

Which of the measures are used in drawing this geometrical figure ?

What are help lines, and how are they drawn ?

When the pattern is finished, what lines bound it?

How many cutting-lines are there in this underwaist pattern?

Which side in all the patterns of this system is used for the back?

How is the front curve of the neck formed?

How is the back curve of the neck formed?

What change is made in the pattern when the waist is larger than the bust?

When the bust is larger than the waist, how is the pattern drafted?

Can this pattern be cut without using parts of inches?

How is this done?

Will the pattern fit well when only whole inches are used?

Take the measures, and draft a pattern.

When the pattern is finished, how is the underwaist cut?

How is this waist put together?

How is it finished?

What measures are taken for a child's underskirt?

How many widths are required for the underskirt of a child from three to five years of age?

How many widths are required for the underskirt of a girl of from five to twelve years of age?

How is an underskirt cut that has a band and shoulder straps?

When it is made with the underwaist, how is the underskirt finished?

How many kinds of darning are there, and what are they?

MATERIALS AND THEIR MANUFACTURE

HOSIERY.

Any fabric which is knitted comes under the head of hosiery. Until 1589 all knitting was done by hand. At that time William Lee, a clergyman born at Woodbridge, England, and a graduate of Cambridge University, invented a knitting machine.

The peculiarity of knitting is that it is weaving with a single thread, and the machinery necessary to accomplish this is most ingenious and complex.

In 1758 Jedediah Strutt adapted the knitting machine to ribbed work, and this was the beginning of a great industry. Until 1816 all machine-knit garments were cut and made like others, with seams. In that year a machine was invented which wove garments and stockings without seams.

One great center of hosiery manufacture is Nottingham, England. There are a large number of manufactories in our own country where very beautiful goods are made. Some idea of the importance of hosiery can be formed from the fact that over 5,000 different articles are made of knitted fabrics.

Cheap hosiery is made on the circular stocking frame. The web is woven in the shape of a tube, and when it is long enough for a stocking it is cut off to form a foot and sewed up by machinery. A machine of this kind makes 1,000 stockings a day.

FELT.

Felt is a kind of cloth which is not formed of woven threads, but is beaten and pressed together. It is used mostly for hats, and is made of wool and the hair of the rabbit, hare, muskrat, and beaver.

Only very fine hair and wool are used for felting, and they must be free from grease, and perfectly clean. After the material to be used is prepared, it is put into a blower with a fan inside which revolves two thousand times a minute. In this way the very fine hair is separated from that which is coarser.

The fine hair or wool for felt hats is weighed, and then fed to a machine which forms it in this way : first two rollers, one with wire teeth and the other with rows of bristles which revolve four thousand times a minute, catch it and send it flying around. As it comes from the feeder it drifts to a copper cone where it clings because the cone is full of holes, and it is over a pit where a fan works so that it draws the air from the outside and the fibers with it. Not a single fiber escapes, and when they are all gathered around, it is sprayed with boiling water which holds it together, so that it can be felted. The felting is a shrinking, pressing process, and when the fabric thus formed is finished and dried it is firm and strong.

When a stiff felt hat is to be made, it is treated to a bath of shellac. Then it is softened and drawn over a wooden block. It is then dipped in the dyeing material. After this it is carefully shaped, and the band and binding are added.

Felt cloth is manufactured in much the same way as felt for hats, only it is made of wool, and is not formed on a cone, but is pressed in flat lengths.

PRINTED FABRICS.

It was in India that the printing of fabrics in various patterns and colors originated. The first printing of fabrics in Europe was near London in 1676.

The processes of printing cloth are very complex, but so perfectly is the machinery for this purpose adjusted that beautiful work is done with great rapidity. At Manchester, England, twenty-five yards of calico are printed in one minute.

There are two kinds of fabric printing, — block printing and machine printing. In block printing the pattern to be printed is cut on a block of sycamore wood as for wood engraving, the parts to make the impression being left prominent, and the rest cut away. An ingenious invention makes it possible to apply several colors at once by means of one block.

The printing which is now almost universally used is by means of cylinders covered with engraved copper. Each cylinder prints a single shade or color. There is a color box in the center of this machine, and by means of screws and other fine mechanical adjustments the pitch of each roller is so arranged that its particular color falls on the proper place with the most minute exactness. Although the machinery for doing this work is intricate and costly, and everything connected with it must be very carefully considered, the process is not expensive. If it were, calico and other printed cloth would not be as cheap as they are.

CHAPTER VIII.

SEVENTH GRADE WORK.

ANY kind of work which is so perfectly planned and executed that there is nothing left to be suggested or desired has reached the dignity of an art, and is a source of much pleasure to the worker. Even the setting of a patch, when properly done, is a pleasant task, and when completed is far from unsightly.

Let the teacher in doing the first work of this grade, which is the gingham patch, call the attention of the pupils to the ease with which the nicest work is done when it is properly planned and each step is carefully executed.

THE GINGHAM PATCH.

The material for this model is a piece of domestic gingham six and one-fourth inches wide and thirteen and one-fourth inches long, figured in quarter-inch checks. From the upper left-hand corner of this gingham model, cut out a piece two and one-half inches long and one and one-half inches wide. Turn in the edges of this opening one-fourth of an inch, taking care to follow a single thread of the fabric. This makes an opening two and three-quarter inches in length and one and three-quarter inches in width.

Cut a gingham patch three and one-fourth inches in length and two and one-fourth inches in width. Crease down one side and one end of this patch one-half inch, and place it under the corner which has been cut out of the ginghan model, in such a way that the edges, when turned in, face each other, and the little squares match to the nicety of a thread. When this has been done, turn back the patch and overhand it to the model, holding it so that the squares of the model are carried out to a thread in the squares of the patch. Cut the corners of that part of the model which is turned in obliquely, trim the seam, and fell the edge of the patch down onto the model with the linen hem.

On the opposite end of the model, measure two inches from each edge, and cut out a piece one and one-half inches wide and two and one-quarter inches long. See that the patch just finished is in the right-hand upper corner, that it may be right side out, and turn in the three edges as for the first patch. Cut a patch three and three-quarter inches long and two and one-quarter inches wide, and turn in one-half

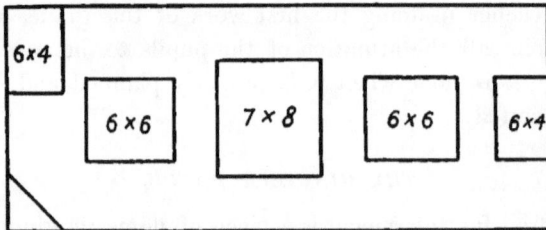

The Gingham Patches.

inch on three sides. Place so that the edges of the model and of the patch, which have been turned in, shall face, and the checks exactly match. Turn back the patch, overhand the four sides; cut the corners of the model, where it is turned in, obliquely, and fell the patch onto the model with the linen hem.

For the next patch, cut out a piece from the model two and one-half inches from the end and two inches from the side, two and three-eighth inches long and two inches wide. After turning in the four sides of this opening, cut a patch one-half an inch larger on each side than the opening, place the patch, after turning in the edges, so that they face and the squares match to a thread, and proceed as in other patches.

When this patch is finished, cut another opening of the same size and form in the opposite end of the model, letting the pupil measure the patch and the work without assistance. Nothing imperfect should be permitted to pass. Work that is not correctly done should be at once ripped and rectified.

For the next patch, an opening is cut in the center of the model two and three-eighth inches long and three and one-fourth inches wide. The patch for this is cut one-half an inch larger on each side than the opening, after it has been turned in one-fourth of an inch on each

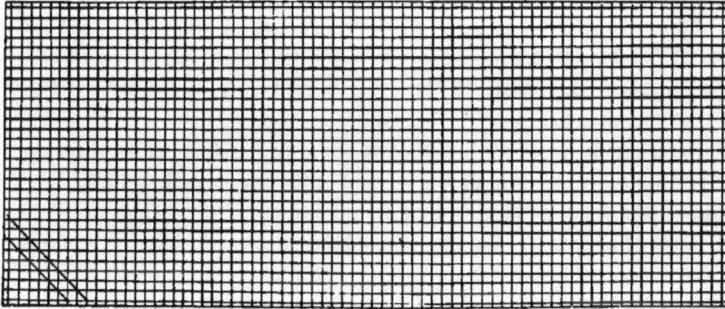

Scheme of Gingham Patches.

of the four sides. The patch is turned in one-half inch on each side, care being taken to turn this as well as all edges, by a thread. Where the edges of the patch and of the opening have been turned in and the patch placed, it is overhanded in on the right side, while on the underside it is simply overcast, as it represents a patch where there is a lining.

The last patch of this model is one that is placed diagonally on the corner. First measure an inch and a half each way from the corner opposite the first patch, and cut diagonally. Cut a patch like this corner with a half-inch added along the diagonal side. Turn in the diagonal edges of the model and the patch, and overhand together; trim the edge of the model that is turned in, and fell the patch with the linen hem. First hem the long sides with an eighth-inch hem, then turn the ends and hem in the same way. The reason for hemming first the sides and then the ends is that the corners may be even and square.

QUESTIONS AND ANSWERS.

What is the first model of the seventh grade ? *Ans.* A piece of domestic gingham figured in quarter-inch checks, six and one-fourth inches wide, and thirteen and one-quarter inches long.

How many patches are there in this model ? *Ans.* Six patches.

What is the first thing to be done ? *Ans.* From the upper left-hand corner of the model, a piece is cut two and one-half inches long and one and one-half inches wide, and the three edges of this opening are turned in one-fourth of an inch.

How large is the patch for this opening? *Ans.* It is three and one-quarter inches in length and two and one-fourth inches in width, and is creased down on two sides so that it fits the opening in the model.

How is this patch placed ? *Ans.* It is so placed that the squares of the gingham match to a thread.

How should the edges of both patch and model be turned? *Ans.* Along the line of a thread.

When the patch has been placed, how should it be sewed ? *Ans.* It should first be overhanded, then the corners of that part of the model which is turned in are cut obliquely, the edge trimmed, and the patch felled down with the linen hem.

Where is the next patch placed ? *Ans.* In the opposite end of the mode. measure two inches from each edge, and cut out a piece one and one-half inches wide and two and one-quarter inches long.

How should the edge of this opening be turned? *Ans.* First see that the patch already placed is in the right-hand upper corner, that it may be right side out, and turn in the edge one-fourth of an inch along the line of a thread.

What size is the patch for this opening ? *Ans.* It is three and three-quarter inches long and two and one-quarter inches wide, and is turned in a half inch on three sides.

How is this patch placed? *Ans.* It is so placed that the squares exactly match, and the edges of the patch and the model are then overhanded together.

How is this patch finished? *Ans.* The corners of that part of the model which is turned in are cut obliquely, the edge is trimmed, and the part of the patch which was creased is felled down over the model with the linen hem.

What is the third patch in this model? *Ans.* The four-sided patch.

Where is the four-sided patch placed? *Ans.* Two and one-half inches from the lower edge of the model, and two inches from the side edge.

How large a space is cut? *Ans.* A piece two and three-eighth inches long and two inches wide.

How large is the patch? *Ans.* After the edges of the opening have been turned in, the patch should be one-half inch larger on each side than the opening.

How is this patch placed? *Ans.* So that the squares match to a thread, and it is then overhanded and felled like the other patches.

What is the next patch? *Ans.* Another like this in the other end of the model.

What is the next work of this model? *Ans.* Cut an opening in the center of the model two and three-eighth inches long and three and one-fourth inches wide, and crease down one-fourth inch on the edges.

How long is the patch for this? *Ans.* One-half inch larger than the opening after the edges are turned.

How is this patch placed? *Ans.* After creasing the edge of the patch along a thread, overhand it onto the model as before. Overcast the under edge instead of felling it, as this represents a patch where there is a lining.

What is the last patch of this model? *Ans.* A diagonal patch on the corner.

How is this patch placed? *Ans.* Measure an inch and a half each way from the corner opposite the first patch, and cut off diagonally. Cut a patch like this corner with half an inch added along the diagonal side; match the squares perfectly; overhand and fell as in the other patches.

Child's Dress.

CHILD'S DRESS.

The next work of this grade is the making of a dress for a child from three to five years of age, or for the big doll, if there is one belonging to the department. The measures for the waist of this little gown are taken as for the underwaist, and the pattern is drafted and cut in the same way.

For the sleeve, the following measures are taken: The length of the arm over the elbow from the shoulder to the wrist; the length from the elbow to the wrist; the length of the inside arm; the length of the inside forearm from the elbow to the wrist; for the width take the size of the arm above the elbow, and add four inches. When these measures have been taken and tabulated, draw a dotted help line parallelogram, using the outside arm length and half the width for the two sides. Measure the length from the wrist to the elbow on line D, and put a point 1. Measure three inches on line A to the left from line D. Draw a curved line from this point 2 to the angle of lines C and D, which gives the outside curve of the sleeve.

For the wrist slant, measure one inch up on line B from the angle of lines B and A, and one inch to the left, point 3, and draw an oblique cutting line from point 2. From point 3, measure the length of the inner arm along line B, point 5. Measure from point 3 along line B the length of the inside forearm, and one inch to the right, point 4. Connect points 3, 4, and 5 with a curved cutting line.

From the angle of lines C and D, measure two inches on line C,

Sleeve Lining.

point 6, and one inch from this, point 7.   Draw a curving line from the angle of lines C and D to point 5 including point 7.

From point 1 on line D measure two inches to the left, and from point 2 at the wrist one inch to the left; connect the two by a curving line with point 6.   From point 6 to point 5 draw a concave line, and the draft of the sleeve is complete.   The under part which is drafted within the upper part can be outlined with a tracer and then cut.   This is, of course, only a plain sleeve lining, but with this as a basis any sort of sleeve desired may be cut.

The skirt of this small frock may be made any length desired, allowing sufficient extra length for a five-inch hem.   Three widths of goods twenty-seven inches wide are required for the little skirt.   This skirt is made with a five-inch placket like the underskirt, but is unlike it in that it is seamed up without being turned to make a French fell.

The waist is made like the underwaist, but the seams are overcast instead of felled.   The sleeve is basted in so that the under-arm seam is an inch and one-half to the front of the side seam of the waist, and it is then sewed in with a backstitch.   The skirt should be divided into four parts, putting two inches more into each of the two back divisions than into the front, and when it is gathered, and the gathers stroked, join it to the waist in the same way as the underskirt, finishing the seam with a narrow instead of a wide facing.   A little ruffle of the material, or of embroidery, or lace, makes a suitable finish for the neck and sleeves.

QUESTIONS AND ANSWERS.

What is the first thing to be done in making a child's dress ?   *Ans.*   To take the measures and draft the waist.

How is this done ?   *Ans.*   The same as with the underwaist.

How many measures are taken for the sleeve ?   *Ans.*   Five.

What are they ?   *Ans.*   The length from the shoulder to the wrist over the elbow; the length from the shoulder to the elbow; the inside length of the arm; the inside forearm; and the width.

What is the first thing to be done in drafting a sleeve? *Ans.* Draw a dotted help line parallelogram with half the width for the short base line A, and the outside arm length for the long perpendicular lines B and D.

How is the outside cutting line of the sleeve found? *Ans.* Measure the length from the wrist to the elbow on line D, and put point 1 an inch to the left; measure four inches on line A to the left from line D, point 2, and draw a curved cutting line connecting it with point 1 and the angle of lines C and D.

How is the wrist slant obtained? *Ans.* One inch on line B from the angle of A and B and one inch to the left, fix point 3, and from it draw an oblique cutting line to point 2.

How is the under-arm seam found? *Ans.* From point 3 measure the length of the inner arm, point 5, and from point 3 the length of the forearm on line B and one inch to the right, point 4, and connect points 3, 4, and 5 with an incurving cutting line.

How is the curved upper part of the sleeve drafted? *Ans.* From the angles of lines C and D, measure two inches on line C, point 6, and one inch up from this, point 7, and connect the angle of lines C and D with points 7 and 5 by a curving line.

How is the under part of the sleeve drafted? *Ans.* From point 1 on line D measure two inches to the left, and from point 2 at the wrist one inch to the left, and connect with point 6 by a curved cutting line.

How is the under part of the sleeve cut? *Ans.* It is traced with a tracer, and then drawn and cut.

How is the waist put together? *Ans.* Like the underwaist except that the seams are not felled.

How is the skirt cut? *Ans.* Three lengths of ordinary width goods as long as is required, with five inches added for a hem.

How is the skirt divided? *Ans.* Into four parts, with two inches more in each of the two back divisions than in front.

How is it finished? *Ans.* It is gathered, and sewed to the waist like the underskirt, but it is finished on the wrong side with a narrow facing.

How is the sleeve put in? *Ans.* The under-arm seam of the sleeve is put an inch and a half toward the front from the side seam of the waist.

BOY'S BLOUSE WAIST AND KILT SKIRT.

For a boy's blouse waist, take measures and draft as for a child's waist, adding four inches to the length, and an inch to half the front, and the same to half the back, at the bottom of the waist, making two inches added to the front for fullness. The additions to the waist for the fullness can be made in cutting by laying the pattern on the doubled goods in such a way as to have a margin of an inch at the bottom beyond the pattern in the front and back. As the blouse is not opened in the back, but the front, the back should be cut on a doubled width of goods, and the inch and three-fourths allowed for the lap should be cut on each side of the front.

This blouse is made like the child's dress waist, with the exception that the bottom is hemmed in a half-inch hem through which an elastic, one inch longer than the waist measure, is run. The added inch on the length of the elastic is for the fastening on either side at the ends.

The underwaist for boys, on which the kilt skirt or trousers are buttoned, is the same as the child's underwaist.

The sleeve of the blouse is cut like the sleeve of the child's dress, with the exception that there is no wrist slant, and the fullness is gathered into a cuff or band.

The kilted skirt is made of straight widths of cloth cut long enough to reach just below the knees, with four inches allowed for a hem. The width is four times the waist measure. The placket is cut and hemmed as in the underskirt. When the skirt has been hemmed, and laid in two-inch plaits, it is felled to a band three inches wide and one inch longer than the waist measure. There are seven buttonholes in the band, — one in each end, two at each side, and one in front.

QUESTIONS AND ANSWERS.

How is a blouse waist drafted? *Ans.* The same as a child's waist, with four inches added to the length, and two inches to the back, and the same to the front for fullness.

Is the blouse waist closed in the back or the front? *Ans.* In the front.

How is the lap provided for? *Ans.* The provision for the front lap is the same as for the back.

How is this done? *Ans.* By adding an inch and three-quarters on each side.

How is the waist finished at the bottom? *Ans.* A half-inch hem is turned, and when it is hemmed, an elastic, one inch longer than the waist measure, is run in, drawn up, and fastened on each side.

How full is a kilt skirt cut? *Ans.* Four times the waist measure.

How long should it be? *Ans.* A little below the knees, with four inches added for a hem.

How is the placket cut? *Ans.* Like the one in the underskirt.

How wide should the kilts be? *Ans.* Usually two inches in width.

How should the band be cut? *Ans.* One inch longer than a loose waist measure, and three inches wide.

Where should the buttonholes be cut? *Ans.* One in the front, and two on each side, with one in each end of the band.

KNEE TROUSERS.

The little trousers which finish the work of this grade are intended to be buttoned onto an underwaist. The measures required are a waist measure, the length of the leg from the waist to the knee, and the inside of the leg to the knee.

A parallelogram with the outside leg measure for two sides, and half the waist measure for the other two, is drawn in dotted help lines. On lines B and D, measure the inside leg measure, and draw a help line E an inch and a half beyond these two lines.

On lines A and C to the left from line D, measure one inch more than half the length of these lines, and between these points draw a cutting line, F. Connect A and E on both sides by an oblique line. Measure an inch and a half down from line A on both sides, and draw a line with half an inch slant for a hem.

Measure up from the angle of lines D and C two and one-half inches,

Boy's Suit.

and connect line E with this point by an oblique line. From the angle of B and C, draw to E a line curved in slightly. One inch and a half from line E on the left side place point 1. Measure up the curved line

Knee Trousers.

of the front two and one-half inches, and place point 2. Cut a fly three inches long, curved on one side to fit the front seam between points 1 and 2, an inch and one-half wide in the center, and narrowed to a point on either side. Face the left front between points 1 and 2, and the seam on the right side of the fly between the same points, pressing the seam open, and stitching it down on either side.

Use two pieces of strong cloth for pockets, ten inches wide and seven inches long. Face these pieces with goods like the trousers on one of the seven-inch sides, so that they face each other. Measure down at the side five and a half inches from the top of each front, notch each side, and turn back and face. Fell on this side of the pocket, measuring seven inches which is not faced, bringing the top of the pocket to the top of the trousers. Care should be taken that the pocket be so basted that the faced part shall be on the inside. When the upper part of the pocket has been felled to the front of the trousers, baste so that the faced part is a little back from the edge, and sew with a backstitch.

Put a fly of material like the trousers, one and one-half inches wide, down five and one-half inches from the top on the sides of the backs. Sew the two fronts together and the two backs. Wet these seams with a sponge, and press until they are flat and perfectly dry. Work a stay at each edge of the front fly. Sew up the side seams as far as the pockets, and press in the same way. Baste so that the back and front seams come together, seam up, stitch, and press. Take a bias piece of black silesia three-quarters of an inch wide; stitch this onto the bottom of the legs, and when the hem has been turned, hem the silesia so that the stitches cannot be seen on the right side. Moisten and press these hems.

Turn in the top of the backs and the fronts, and fell on an inch and a half bias facing. Cut a double band an inch and a half wide. Put a buttonhole in the center of this band, and another half-way between the center and the sides of the trousers. Double in these bands, and stitch them onto the trousers at the top, with the edge of the band a little below the edge of the trousers. Fasten at the ends and between each buttonhole. Fasten the fly on the back of the trousers to the front, and work a stay where the two edges come together. Put a buttonhole on each side of the front of the trousers, and a button on the back so that the two edges come together.

To remove the shiny look caused by pressing, wet a cotton cloth, and, after wringing it quite dry, lay it over the shiny part, and go over it with a hot flatiron very lightly. Remove the cloth quickly, and brush. In doing this the iron should be kept in the hand, and not set down on the cloth.

QUESTIONS AND ANSWERS.

What are the measures taken for boys' knee trousers? *Ans.* A waist measure, the length of the leg from the waist to the knee, and the inside of the leg to the knee.

What is the first thing to be done in drafting boys' trousers? *Ans.* Draw a dotted help line parallelogram with the outside leg measure for two sides, and half the waist measure for the other two.

What is the next step? *Ans.* On lines B and D mark off the inside leg measure, and draw a help line E an inch and a half beyond these two lines.

How is the line that divides the back from the front found? *Ans.* On lines A and C to the left from line D, measure one inch more than half the length of these lines, and connect by the cutting line F.

How is the lower part of the trousers leg formed? *Ans.* On both sides connect A and E by an oblique cutting line.

How is the hem provided for? *Ans.* Measure an inch and a half down from line A on both sides, and draw a line with half an inch slant for a hem.

How is the extra length in the back obtained? *Ans.* From the angle of lines D and C measure up two and one-half inches, and connect this point with line E by an oblique cutting line.

What is the next thing to be done? *Ans.* From the angle of B and C, draw a line to E curving in slightly.

Where is the opening for the front fly? *Ans.* One and one-half inches from line E is point 1, measure up the curved line two and one-half inches, point 2, — between these points is the space for the fly.

How is the fly cut and placed? *Ans.* It is three inches long, and curved to fit the front seam on one side, and is circular on the other.

Where is this placed? *Ans.* It is seamed on the right side, pressed and stitched, and the left side is faced between points 1 and 2.

How are the pockets cut? *Ans.* Ten inches wide and seven inches long.

How are the pockets prepared? *Ans.* They are faced on two sides with material like the trousers.

How are they put in? *Ans.* The sides which are not faced are felled to the fronts; they are then brought together and sewed, the tops being basted to the top of the trousers, and afterwards sewed on with a band.

How is the back part of the trousers opposite the pockets finished? *Ans.* With a fly five and one-half inches long, and one and one-half inches wide.

What is the next thing to be done? *Ans.* Sew the two fronts together on either side of the fly, and after sewing the backs together and thoroughly pressing the seams, sew and press the side seams.

What is the last seam sewed? *Ans.* The inner leg seam, which should be sewed from the center with the two seams evenly opposite each other.

How should the seams be pressed? *Ans.* They should be wet with a sponge, and pressed until perfectly dry.

How is the front fly finished? *Ans.* A stay is worked at each side.

How are the sides finished? *Ans.* With a stay like the front fly where the seams begin.

How is the bottom of each leg finished? *Ans.* A bias strip of silesia three-fourths of an inch wide is stitched to the bottom, the hem is then turned, and the edge of the silesia turned and hemmed.

How is the top finished? *Ans.* Turn in the top of the back and front, and fell on a bias facing an inch and a half wide. Put a buttonhole a quarter of an inch from the edge, and a half-inch from the top in the front, and set buttons in the back so that the two edges come together.

How are the trousers buttoned onto the underwaist? *Ans.* Cut a band one inch shorter than the width of the fronts, and three inches wide, and another the same width one inch shorter than the width of the back. Double in the edges, and make one buttonhole in the center of each, and another in each side half-way between it and the edge of the trousers. Stitch this band on so that it does not show above the tops of the trousers, and fasten between each buttonhole.

How are the trousers finished ? *Ans.* After the hems and the top have been thoroughly pressed, sponge by wetting a cloth, and, after wringing it quite dry, lay it on the seams on the right side, touch lightly with a heated flatiron, and brush quickly ; in this way the shine caused by the pressing is removed.

QUESTIONS FOR REVIEW.

What is the first model of the seventh grade ?

How many patches are there in this model ?

As a rule, how much larger is a patch than the opening to be mended ?

Is it necessary that the figure of the goods be matched ?

When the edges have been turned, how is the patch sewed into the opening ?

What is the second work undertaken in this grade ?

How many measures are taken for the sleeve ?

When the measures have been taken, how is the sleeve pattern drafted ?

How is the under part of the sleeve cut ?

How is the skirt to the child's dress cut ?

How is the sleeve basted into the waist ?

How is the skirt divided before it is gathered ?

How is the boy's blouse waist cut ?

Is the boy's blouse closed in the front or back ?

How full and how long is a boy's kilt skirt ?

What are the measures taken for boys' trousers ?

How is the pattern for boys' trousers drafted ?

How is the fly for the front cut and placed ?

How are the pockets cut ?

How are the pockets put in ?

How is the back part of the trousers opposite the pockets finished ?

In what order are the seams of the trousers sewed ?

How is the top of the trousers finished ?

How are the bottoms finished ?

How is pressing done ?

After the trousers have been thoroughly pressed, how are they finished ?

How is sponging done ?

MATERIALS AND THEIR MANUFACTURE.

NEEDLES.

DID you ever consider how much work it must be to make a needle? Each one must be absolutely perfect or it would be utterly useless. And what a fine, delicate little instrument it is; very different indeed from the first needles used by mankind, which were made of fish bones.

In the first place, only the best steel wire can be used for needles; and this wire comes to the needle factory in great coils, and is cut with big shears into lengths sufficient for two needles. ·When these have been straightened, several thousands of them are packed into strong iron rings. These are heated red-hot, and then pressed onto an iron plate having two grooves in which the rings run. Constantly pressed by a slightly curved tool, back and forth they go until all the wires become perfectly even and straight.

The next thing to be done is to point both ends of these wires on a dry grindstone that revolves very fast indeed. There is a sort of hood over this flying stone to keep the steel dust away from the person who does this work, and a strong current of air helps to draw it away. Still there is so much of the fine steel dust all about, that some of it is breathed into the lungs, and the result is that the workers soon become ill, and it is necessary to secure others to take their places, so that the making of needles costs many lives each year. There has been a machine invented to do this work which does the grinding very rapidly, but not quite as well as it is done by hand.

When the needles have been ground, a groove is stamped in the center for the two eyes, for it must be remembered that each wire represents two needles. Through these stamped heads, the eye for each needle is punctured. Now the wire has become two needles, held together by a thin bit of steel. One hundred of these double needles are threaded onto two fine wires and clamped tightly together; the needles are then broken apart so that the head of each one can be rounded off with a file.

After the heads of the needles are rounded off, they are heated red-hot and plunged into an oil bath, and then once more heated. When they have cooled, they are put into bundles of several thousands each, are mixed with soft soap, oil, and emery powder, and tied up in canvas covers. They are then put into a machine that rolls them backward and forward until they are well scoured. When they have been taken out of the covers and washed, they are put into others containing putty-powder instead of emery. After this polishing process, the needles are unpacked, washed in an alkaline solution, and dried in sawdust. They are then put into trays, and are made parallel by a jerking motion. After this they are brought into one direction by a "header," who has a thick cushion on his finger into which he presses a large number of needles.

After the imperfect needles have been thrown out, the heads are blued by heating in a flame of gas. When this has been done, the needles are strung on a rough steel wire, over which is spread a fine paste of oil and emery, and are moved backward and forward until the eyes are perfectly smooth. After a final polishing on a rapidly revolving buff-wheel, the needles are assorted, put into papers, and are then ready for use.

EMERY.

It would not be easy to tell from what part of the world come the fine, irregular, sharp crystals that make your needle so smooth when you run it through your emery bag.

Perhaps this emery has been a great traveler, and come all the way from Cape Emerie, on the island of Naxos, in the Ægean Sea, where the best emery in the world is found, and from which it takes its name. It is more than likely to have come from this island, as there are many tons shipped from there to all parts of the world every year. It may have come, however, from Sweden, Saxony, Spain, Greenland, or Massachusetts; for emery is found in all these places. Wherever it came from, it is a sort of sapphire, and was in the beginning bluish or brownish gray in color, although it is often artificially colored a rich reddish brown.

Emery is first crushed with steel stamps; then it is sifted. It is used in cutting marble and granite, also for polishing plate-glass, crystal, metals, and gems. as well as needles.

PINS.

Although such pins as we use now, for so many different purposes that it would be very difficult to enumerate them, are of a comparatively recent date, pins of some kind seem always to have been used. The first pins were thorns, and even at the present time the peasant women of Upper Egypt use these to fasten their dresses.

The pins now in common use are made very rapidly and almost entirely by machinery. After the wire of which they are made has been wound on a reel, it is passed between straightening pins set in a table. When a pin has passed through these straightening pins, it is caught by lateral jaws, beyond which enough of the end projects to form a pinhead; against this projecting portion a steel punch is thrown, which compresses the metal by a die arrangement into a head. The pin length is immediately cut off, and drops into a slit which lets the wire pass through, but retains the head so that the points are held against a filecut revolving steel roller. The pins are carried along this roller by gravitation, until they fall out at the extremity, well-pointed pins.

The pins are next cleaned by being boiled in weak beer, and are then arranged in a copper pan in layers alternating with layers of grained tin. A sprinkle of argol and water enough to cover the pins is added, and the whole is boiled for several hours, after which they come out having a silvery appearance.

After being washed, they are dried by revolving in a big vat with dry bran. The finished pins are stuck in papers by means of an auto-matic machine which also folds the papers. The pins are then ready for the market.

Pins were a very different article during the reign of Henry VIII. from what they are at the present time. A law was enacted then that "No person shall put on sale any pins as shall not be doubled-headed and soldered fast to the shank, well smoothed, shaven, filled, canted, and sharpened." It was during the reign of Charles I. that a pin-makers' corporation was first founded in London.

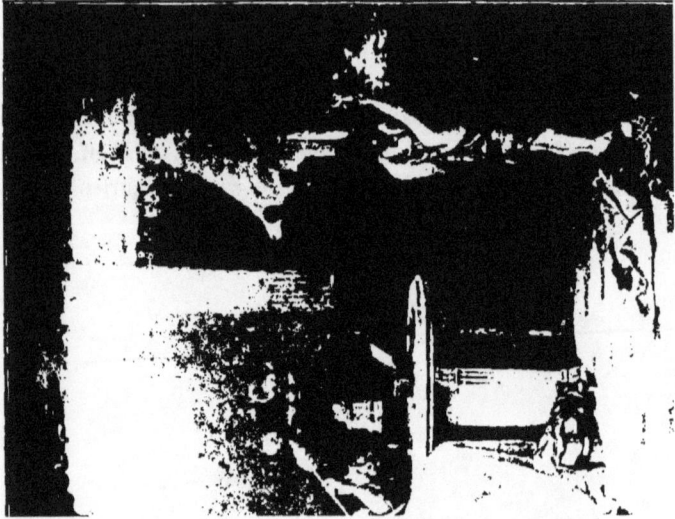

CHAPTER IX.

EIGHTH GRADE WORK

THE principal work of the last grade in this system of sewing and garment cutting is the drafting, cutting, and making of an infant's outfit and of a dress for a young girl. The latter is usually the graduating dress of the pupil, and is of Victoria lawn or some other fine white goods. This gown is in every way perfectly simple, and involves only the basic principles of dressmaking. More than this would open a field quite beyond the scope of the present work, which, as has been stated, is, first of all, educational; its chief purpose being to make the pupil so thoroughly mistress of her mind and hand that she is able to undertake with ease and with success any of the various branches of needlework, such as tailoring, dressmaking, fine sewing, or art work.

Since one of the foundation principles of the system is exactness and thoroughness, it has not been deemed advisable to introduce fancy work of any sort, as it would be quite impossible to give adequate instruction in this or any other lines of advance needlework in a text-book of this kind. It is nevertheless true, that a pupil who has taken the entire course indicated in this book will have become so complete a mistress of the needle and of the fundamental principles governing its use, that the technicalities of any particular line can be easily and quickly mastered. More than this, as the pupil has learned, in each instance, to combine and to separate, it will be easy for her to differentiate results indefinitely. This is demonstrated in the miss's waist of this grade, which is simply the straight, curveless garment of a child, transformed by slight changes into one suited to a developed girl.

While the graduating dress and the infant's outfit form the principal work of this grade, the first work is the linen patch. This model is the finest and most difficult needlework in the course; and when it can be executed with neatness the pupil is mistress of the needle, and can with care readily acquire the technicalities of any special department of needlework.

THE LINEN PATCH.

For this model, take a piece of rather fine linen six and one-half inches wide and seventeen inches long. Three and one-fourth inches from each end draw two threads, and turn a hem for hemstitching.

Crease down one-fourth of an inch of the model on each side, and turn under one-eighth of an inch for the hem.

When these two hems have been basted, measure an inch and a half from the edge of the long side and the same distance from the drawn threads at the end, mark off a square of three inches, draw a thread, and cut. After creasing down the edge of this opening one-fourth of an inch, fold in such a way as to bring the four corners of the space together, and crease. Then turn and fold in an opposite direction; again bring the corners together, and crease. In this way the center of each side of the space to be patched is obtained.

The opening after it has been creased is three and one-half inches square. Cut a patch four and one-half inches square, fold, and crease

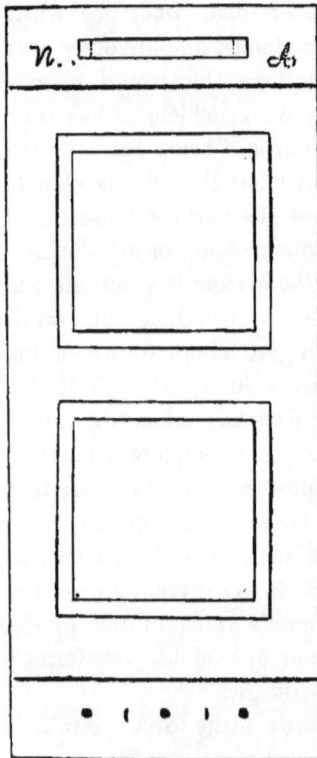

Pattern of Linen Patches.

through the center in both directions. Turn down the edge of the
patch one-half an inch all around, place so that the creases in the patch
and those in the opening come together, and overhand the patch into
the opening. Turn on the wrong side, and cut the corners of the model
diagonally; trim these edges, then turn the edge of the patch one-
fourth of an inch, and fell onto the model with
a linen hem.

Set a second patch in the opposite end of
the model in the same way. When this patch
is finished, hemstitch the ends, and hem the
sides with the linen hem. The reason the hemming is left until the last is, that should there
be a mistake in the patches, the labor of hemming will not be lost.

Take a piece of linen tape three and one-
half inches long, and one-fourth of an inch
wide, and baste in the center of the hem at the
right hand end, one and one-half inches from
the edge on each side. Turn one-fourth of an
inch at each end, and backstitch one-quarter of
an inch from the end, and after it has been
turned in hem around the edge and end. This
hemming with the backstitching forms a square
at each end of the tape loop. At even dis-
tances from the edge of the hem, mark a square

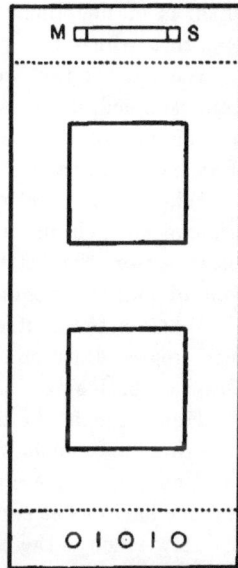

Model of Linen Patches Finished.

with the point of the stiletto three-fourths of an inch on each side of
the tape loop, and within this space have the pupil put the initials
of her name.

On the other end of the model, one and one-half inches from the
edge and the center of the hem, make an eyelet with a stiletto, and
work with a blanket stitch. Measure three inches to the right and
place a second eyelet, and put another half-way between the two. In

the center of the hem, three-quarters of an inch from the first eyelet, put a loop of five threads, covered with buttonhole stitch, and the same distance from the second eyelet, put a second loop, which finishes the model.

QUESTIONS AND ANSWERS.

What is the first model of the eighth grade? *Ans.* The linen patch, which is of fine linen six and one-half inches wide and seventeen inches long, into which two patches are set.

What is the first work to be done? *Ans.* Three and one-fourth inches from each end, draw two threads, and turn and baste a hem. Crease down on each side of the model one-fourth of an inch, and turn under one-eighth of an inch, and baste.

Where is the first patch set? *Ans.* When the hems have been basted, measure an inch and a half from the edge of the long side, and the same distance from the drawn threads at the end, and after marking off with the point of a stiletto a square of three and one-half inches, draw threads and cut.

What is the next step? *Ans.* When the edges of this opening have been creased down on all sides one-fourth of an inch, fold, and crease in the center on both sides.

How large is the opening after the edges have been turned in? *Ans.* It is three and one-half inches square.

How large a piece of linen is to patch this opening? *Ans.* A piece four and one-half inches square.

How much is the edge of the patch turned in? *Ans.* One-half an inch on each side.

How is this patch placed? *Ans.* It is first creased through the center both ways, and the creases of the patch are placed even with the corresponding creases in the model, and the two are overhanded together.

How is the patch finished? *Ans.* The edges of the model which have been turned in are cut diagonally at the corners, then trimmed, and the edges of the patch turned in one-fourth of an inch, and felled to the model with the linen hem.

How is the second patch placed? *Ans.* In the opposite end of the model, in the same way.

How are the edges finished ? *Ans.* The ends are hemstitched, and the sides hemmed with the linen hem.

What is placed at the right-hand end of the model on the broad hem ? *Ans.* A loop and two initials.

How is the loop cut and placed ? *Ans.* It is a piece of linen tape three and one-half inches long and one-fourth of an inch wide, put on in the center of the hem one-half an inch from the edge. It is turned in a quarter of an inch, backstitched a fourth of an inch from each end, and hemmed at the ends in such a way that with the backstitching it forms a square.

How are the initials placed ? *Ans.* A square of three-fourths of an inch is drawn in the center of the hem on either side of the loop, with the point of the stiletto, and in each of these squares an initial is placed.

What finishes the other end of the model ? *Ans.* Eyelets and loops.

How are the eyelets placed ? *Ans.* One and one-half inches from the edge of the model in the center of the hem, make an eyelet with a stiletto, and work it with the blanket stitch; measure three inches to the right, and make another eyelet, and place still another exactly half-way between these two.

How are the loops placed ? *Ans.* In the center of the hem between the first and second, and between the second and third eyelets.

INFANT'S OUTFIT.

Up to this point the sewing has been done by hand; in making the infant's outfit, machine sewing is first introduced. While the class makes the entire outfit, the different pieces are made by individual pupils, as there is not sufficient time for each pupil to make the entire set. Each garment, however, should be so carefully explained to the class that no member of it need have any difficulty in making the whole wardrobe.

The outfit consists of a flannel band, a shirt, a pinning blanket, a flannel skirt, a cambric skirt, a wrapper, a sack, and a dress. There is also a miniature bed, furnished with sheets which are hemstitched at the top, pillow slips, a blanket and quilt, all made to fit the little bed.

THE FLANNEL BAND.

This band should be twenty-four inches long and nine inches wide. Turn down one inch on the two opposite sides and the same on the two ends. The hem should be turned in this way that the corners may be neat and also uniform. When the hem is turned and basted, catch-stitch it on the right side that the smooth surface may come next the body.

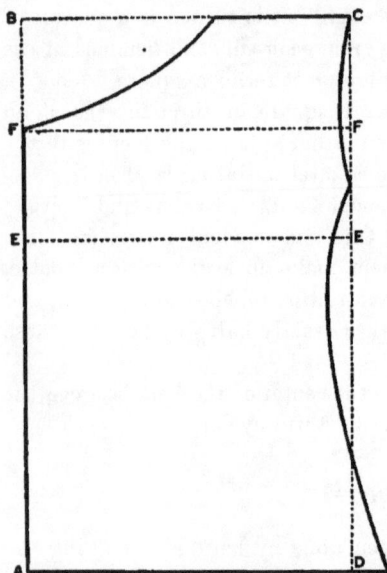

THE SHIRT.

The material of which this shirt is made is fine linen. The measures used are twenty-four inches for the bust measure; eight inches for the front length, with two inches added for the neck; six inches for the side length; two and one-half inches for the shoulder. The shoulder is measured from the point where the lines C and D intersect, and from this point draw a curved line to the intersection of lines B and F. A slight curve is drawn along line D to one inch below F. This forms the armhole. To the right of line A, add one-half an inch, and from this point to a point a little below line F draw a slightly curved line for the under-arm seam. The pattern should be so placed that both the front and the shoulder are on a doubled fold of the goods. This is accomplished by doubling the goods across the width for the shoulders, and lengthwise for the front, making the goods of four thicknesses. Through the lace trimming about the neck, a ribbon is run, and this is drawn up and tied.

Infant's Shirt.

Cut two thirty-six-inch lengths of flannel which is thirty-six inches wide, and seam the two together. Turn a two-inch hem on the two sides and across the bottom, and catch-stitch on the right side. The waist of this little open skirt, which is like the other waists of the system, with only this difference, that two inches are added to the front length instead of two and one-half, is drafted from the following measurements : Bust measure twenty-four inches, waist measure twenty-four inches, front length six inches, side length four inches, shoulder two and one-half inches. In drafting this little waist, an extra help line is drawn one inch below line C to make a straight shoulder. As the front and back are very similar, they are not drafted separately.

For the front, lay the pattern on the doubled goods, pin and cut, allowing one-half an inch for seams at the sides and

Underwaist.

shoulders, and one and one-third inches for the hem, and closing at the right of the pattern. The back is cut in the same way, allowing one-half an inch for seams at the sides and shoulders, but leaving an opening. The armholes and neck are cut out a little more in the front than in the back. An inch is cut out around the neck, and it, together with the armholes, is finished with a half-inch bias binding of the goods; the side

and shoulder seams are finished with a French fell. The pinning blanket is plaited to fit the twenty-four-inch waist measure of this underwaist, and is joined to it, the seam being finished with a piece of bias
cambric. The buttonholes of this waist are two inches apart.

THE FLANNEL SKIRT.

Cut two lengths of thirty-six-inch flannel thirty-one inches long.
Join both sides, and catch-stitch the seams. Turn a two-inch hem at
the bottom, and catch-stitch. In the center of one width, cut a six-inch
placket. Turn a hem on the right side of this placket one-half an inch,
and on the left one-fourth of an inch wide, and catch-stitch. This hem
is stayed on the right side by backstitching in the shape of a right angle,
and on the wrong side a bar is worked. This skirt is plaited, and joined,
as the pinning blanket is, to an underwaist, which is opened, not in the
front, but in the back. With this exception it is like the one used with
the pinning blanket.

THE CAMBRIC SKIRT.

This skirt is of two lengths of thirty-six-inch cambric thirty-four
inches long. After these lengths are joined on both sides, a five-inch hem
is turned and stitched, and the bottom is finished with lace or embroidery.
If a ruffle is desired, one-half the width of the skirt should be allowed for
fullness. The skirt is gathered, and of course the gathers are carefully
stroked; and it is then sewed to an underwaist, with a bias piece to fell
over the seam. The placket is like that of the flannel skirt.

THE WRAPPER.

The same measures are used in drafting the wrapper which are used
for the waist, except that the drawing is extended twenty-eight inches
beyond the waist, and one and one-half times the width is added for the
slant, which in this case is nine inches, as the width is six inches, and

the slant begins at the armhole, and extends to the bottom of the garment. The bottom of the garment is curved from the center of the width to the side seams, from which two inches are taken, as the slant

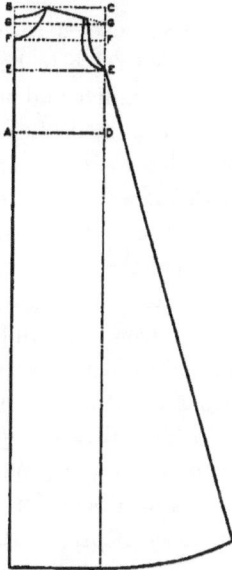

Infant's Wrapper.

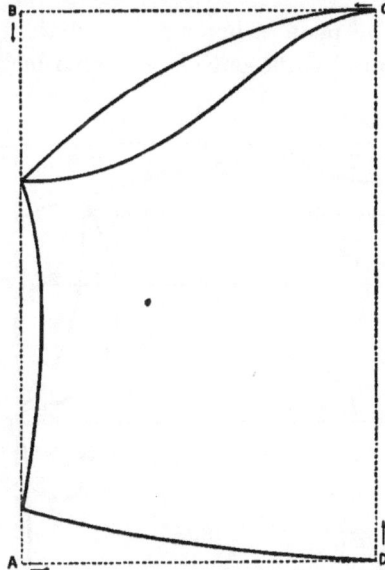

Bishop Sleeve.

makes the seams longer than the rest of the garment. The pattern is placed on the doubled cloth, with one and three-fourths inches allowed for the front lap. There is the same allowance for seams as in the waist. The material used is either flannel or cashmere, and if the material is heavy the seams should be clipped.

For the collar, cut a piece of the goods, five inches wide, the size and shape of the neck. After shaping it, join the outer edges on the wrong side.   When finished, this collar should be two and one-half inches wide.   Pin the center of the collar to the center of the garment in the back, and sew on the upper side, felling the under side over the seam.

The plain bishop sleeve is drafted like the dress sleeve, with three measures, — the outside arm nine inches, the width twelve inches, and the inside arm five and one-half inches.   The band at the hand is six inches long, and two and one-half inches wide.   In putting the sleeve in the armhole, the seam of the sleeve is placed one and one-half inches toward the front from the under-arm seam, and most of the fullness is gathered about the shoulder seam.   The little wrapper is finished down the front with ties of baby ribbon, or it may be buttoned if preferred.

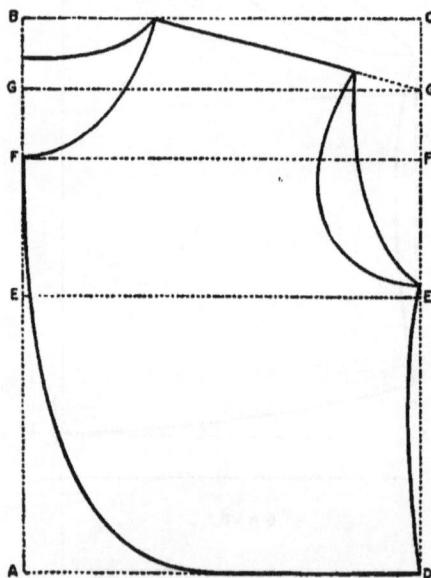

THE SACK.

The waist pattern, with one inch added to each of the side seams in excess of the allowance for seams, and one and one-half inches added to the length, with a slight curve below the line E, forms the sack.   The collar and sleeves are like those used for the wrapper.   A plain sleeve may be used if desired.   This sack, which is of some soft woolen material, is either pinked or finished with an embroidered scallop about the edge.

Infant's Sack.

THE DRESS.

Whatever the style of the dress, it should measure one yard from the neck to the lower edge of the hem. If it is a dress with a waist, the skirt is similar to the cambric skirt, and the waist identical with the one already described. The bishop sleeve, like those in the wrapper, or a plain sleeve, may be used. If the dress is a yoke with full skirt, the yoke is simply the waist cut off so that the desired width is left, with the skirt cut about four inches longer, or whatever the difference is between the width of the yoke and the entire waist. With the exception of the length of the skirt, the yoke dress is cut the same as one made with the waist.

QUESTIONS AND ANSWERS.

What constitutes a simple outfit for an infant ? *Ans.* A flannel band, a pinning blanket, a flannel skirt, a cambric skirt, a wrapper, a sack, and a dress. There should also be a little bed with sheets, pillow cases, a blanket, and a quilt.

What are the measures for the flannel band ? *Ans.* It is cut twenty-four inches long and nine inches wide.

How is it finished ? *Ans.* With an inch-wide hem, turned first along the sides, and then across the ends, and is catch-stitched on the right side.

Why is the hem turned this way ? *Ans.* The two sides and then the two ends are turned that the corners may be neat and uniform, and the hem is turned on the outside that the smooth side may come next the body.

Of what material is the skirt ? *Ans.* Fine linen.

How many measures are used, and what are they ? *Ans.* Four, — a bust, a front, a side, and a shoulder measure.

How is the pattern for the shirt laid on the goods ? *Ans.* So that the shoulders and also the front are on a doubled fold of the goods.

How is it drawn up about the neck ? *Ans.* With a ribbon run through the lace trimming.

What is a pinning blanket ? *Ans.* It is an open flannel skirt of two lengths of thirty-six-inch flannel, thirty-six inches long.

How is it made? *Ans.* The two lengths of flannel are seamed together; a two-inch hem is turned down the sides and across the bottom; it is then plaited, and sewed on an underwaist.

How is the underwaist for the pinning blanket drafted? *Ans.* Like the other waist of this system, with two inches added to the front length, instead of two and one-half.

What are the measures used for this infant's waist? *Ans.* A bust measure of twenty-four inches, waist measure twenty-four inches, front length six inches, side length four inches, shoulder two and one-half inches.

Is the front different from the back? *Ans.* No, it is the same, except that the neck and armholes are cut out a little more in the front than in the back.

How is the waist cut? *Ans.* The pattern is laid on the doubled goods, one-half an inch allowed for seams, and one and three-fourths inches for hem and closing.

How is the pinning blanket joined to the waist? *Ans.* It is plaited, and then sewed onto the waist with a narrow bias band of cambric to fell over the seam.

How is the flannel skirt cut? *Ans.* Of two lengths of thirty-six-inch flannel, thirty-one inches long.

How is it made? *Ans.* The lengths of flannel are seamed together, a hem two inches wide is turned, a six-inch placket is cut in the center of one width; and after this is hemmed and finished, the skirt is plaited, and joined to the underwaist as the pinning blanket is.

How is the cambric skirt cut? *Ans.* It is cut of two lengths of thirty-six-inch cambric, thirty-four inches long.

How is it made? *Ans.* Like the flannel skirt, except that there is a five-inch hem turned, and it is gathered instead of being plaited.

How is the wrapper drafted? *Ans.* Like the waist, with the drawing extended twenty-eight inches beyond the waist line, and one and one-half times the width added for the slant.

How is the garment cut? *Ans.* The pattern is placed on the doubled cloth, with one and one-fourth inches allowed for the lap in front.

What are the measures for the bishop sleeve of this wrapper? *Ans.*

For the outside arm nine inches, the width twelve inches, and the inside arm five and one-half inches.

How is the sleeve put in the armhole? *Ans.* The seam of the sleeve is placed one and one-half inches toward the front from the under-arm seam.

How is the sack drafted? *Ans.* Like the waist, with one inch added to each of the side seams in excess of the allowance for seams, with one-half an inch added to the length, and a slight curve below the waist line E.

What kind of a sleeve has the sack? *Ans.* Either a loose coat sleeve or a bishop sleeve like the wrapper.

How is the edge finished? *Ans.* Either with pinking, or an embroidered scallop.

How long should the dress be from neck to hem? *Ans.* One yard.

How is it drafted? *Ans.* Like the waist and cambric skirt.

How is the dress with yoke drafted? *Ans.* The waist pattern is cut off, leaving as much of the upper part as is desired for a yoke. The skirt is cut about four inches longer than the regular dress skirt, or still longer if the yoke is very short.

GIRL'S WAIST.

Take the measures as for the child's waist. Then draw a parallelogram as for that waist, with half the bust measure for the base line A, and the front length with two and one-half inches added for the vertical line B, which is drawn as a dotted help line. The horizontal line C is also a dotted help line, while the vertical line D is a cutting line. Measure the side length on the vertical lines B and D, and from these two points draw a dotted help line E. Measure the front length on the vertical lines B and D, and draw a dotted help line F. Measure one-fourth of the bust measure on line A from the left-hand lower right angle of the parallelogram, and also on line C, and draw a straight dotted help line G.

Take half the back width, and measure on the base line A from the lower right-hand angle of the parallelogram, and draw the vertical help line H. From line C down line D, measure an eighth of an inch, point 1, and one inch and a half on line C and half an inch up, point 2, and

connect with a slightly curved line which forms the back neck. From
point 2 to the intersection of lines H and F, draw an oblique line ; measure the length of the shoulder, point 3. Draw a slightly curved line from
this point to the intersection of lines E and H for the back arm scye.

Measure off one inch on line A from the angle of A and D, and draw
a help line from the back of the neck to this point. Measure half

Girl's Waist.

the shoulder length, point 4. Measure three inches to the left on line
A from the angle of lines A and D, and draw a dotted help line I to
point 4. Measure three inches along the arm scye from the end of the
shoulder line, point 5, and draw a slightly curving line to the help line
I, and follow it to line A. This line is the back form.

Add an inch and three-quarters to lines A and C for the back lap,
and connect these two points with the cutting line J.

Girl's Graduating Dress.

Measure two inches and a half down line B, point 6, and the same distance on line C, point 7, and connect with a curved line, which forms the neck. From this curved line to line A, make B, the dotted help line, a cutting line. From the neck line on C, draw an oblique line to the intersection of lines H and F, point 8. Measure the length of the shoulder from the neck on this line, point 9, and draw a curved line to the intersection of E and H, forming the front arm scye.

Measure from the angle formed by lines A and B, one inch and a half to the right, point 10; again measure one inch and a half from point 10 to point 11. Half-way between these two, draw a straight help line upward five inches, and connect points 10 and 11 with an oblique cutting line coming together at the top of this five-inch help line. This forms the first dart. Measure one inch to the right of the last line, point 12, and an inch and a half to the right, point 13. Half-way between these two points, draw a straight dotted help line upward six inches. Connect points 12 and 13 with the upper part of the six-inch dotted help line by oblique lines, which gives the second dart.

After excluding the darts, if line A is longer than half the waist measure, take off the difference equally on each side of line H.

WAIST OF MANILLA PAPER.

When the measures have been taken, let them be reduced to quarter inches. Lay the pattern on the manilla paper in such a way that the front is on a fold of the paper, allowing an inch at the back for buttons and buttonholes, and an eighth of an inch for the shoulder and side seams. Sew the seams with a backstitch with a No. 8 needle and No. 50 white thread.

If the difference between the bust and waist is four inches, there should be but one dart and no slant at the back; if there is a difference of five inches, there should be one dart and one inch slant at the back; if there is a difference of seven inches, there should be two darts and one inch slant at the back.

QUESTIONS AND ANSWERS.

How should the measure for a girl's waist be taken? *Ans.* The same as for a child's waist.

When the measures are taken, what is done? *Ans.* A parallelogram is drawn with half the bust measure for the base line A, and the front length with two and a half inches added for the vertical line B.

What are the next lines? *Ans.* Measure the side length on vertical lines B and D, and from these two points draw a dotted help line E; measure the front length on vertical lines B and D, and draw a dotted help line F.

How is the next line found? *Ans.* Measure one-fourth of the bust measure on line A from the left-hand lower right angle of the parallelogram, and also on line C, and draw a straight dotted help line G.

How is the line that forms the under-arm seam found? *Ans.* Measure half the back width on the base line A from the right-hand angle of the parallelogram, and draw perpendicular help line H to C.

How is the curve in the back of the neck formed? *Ans.* From line C, down line D, measure an eighth of an inch, point 1, and one inch and a half on line C and half an inch up, point 2, and connect the two points with a slightly curving line.

How is the shoulder slant formed? *Ans.* From point 2 to the intersection of lines H and F, draw an oblique line, and measure the length of the shoulder, point 3.

How is the back arm scye formed? *Ans.* Draw a slightly curving line from point 3 to the intersection of lines E and H.

How is the slant of the back formed? *Ans.* From the angle of A and D, measure an inch on line A, and draw a help line from the back of the neck to this point.

How is the side form obtained? *Ans.* Measure half the shoulder length, point 4, then measure on line A three inches to the left from the angle of lines A and D, and draw a dotted help line I to point 4. Measure three inches along the arm scye from the end of the shoulder line, point 5, and draw a slightly curving line to help line I, and follow it to base line A.

How is the back lap formed? *Ans.* Add an inch and three-quarters to lines A and C, and connect with cutting line J.

How is the front part of the neck formed ? *Ans.* Measure two inches and a half down line B, point 6, and the same distance on line C, point 7, and connect with a curved line.

How is the front shoulder formed ? *Ans.* From the neck line on C, draw an oblique line to the intersection of lines H and F, point 8, and measure the length of the shoulder on this line point 9.

How is the front arm scye formed ? *Ans.* From point 9, draw a curved line to the intersection of lines H and E.

How is the first dart formed ? *Ans.* From the angle of lines A and B, measure one inch and a half to the right, point 10; again measure one and a half inches to the right, point 11; half-way between these, draw a help line straight upward, five inches in length, and connect points 10 and 11 with the top of this help line by oblique lines.

How is the second dart formed ? *Ans.* Measure one inch to the right of the last dart, point 12, and again an inch and a half to the right, point 13.   Half-way between these two points, draw straight upward a dotted help line six inches in length; connect points 12 and 13 with the upper part of this line by oblique lines.

If, after the darts are taken out, line A is longer than half the waist measure, what is done ? *Ans.* Half of the difference is taken off from each side of line H.

THE SLEEVE, THE SKIRT, AND REVIEW WORK.

The sleeve is measured, drafted, and cut on the same plan as the sleeve of the child's waist.   As the pupil is now familiar with the principles of the system, it is an easy matter to make such alterations in the sleeve as prevailing styles may demand.

Draft the sleeve like the lining described above to point 5, which is connected with the right angle formed by lines C and D by a curved line.   The under-arm curve is identical with that of the lining.   From point 2 measure two inches toward the wrist, and draw a dotted line to point 1.   In this pattern line D is made a cutting line from C to point 1, and points 1 and 2 are connected by a cutting line.

In cutting the sleeve, line D is placed on a fold of the goods, and is cut along the solid line, allowing one-half an inch for the inner seam. Be careful in cutting a pair of sleeves that they are not both cut for one arm. For practice, the measures should be reduced to one-fourth of an inch, and several pairs of sleeves of manilla paper cut and made.

The skirt is of the plain, full sort, made of straight widths, with five inches added to the length desired, for the hem. The placket is like the child's skirt placket. Before gathering, the skirt should be divided into four parts, with six inches more in each. of the two back divisions than in each of the two front parts.

The last work of this grade and of the system is a review of all the work, beginning with the practical darning, and closing with a complete outfit, either for the big doll or a small child. The pupils should have, when they finish the course, a complete set of the models of the system which have been perfectly executed by them.

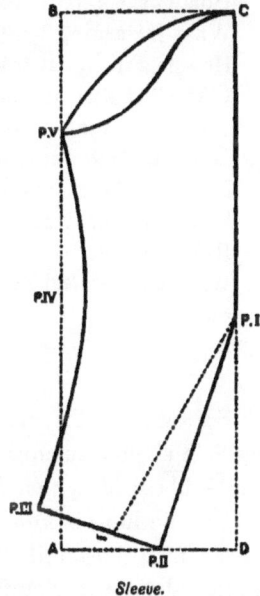

Sleeve.

QUESTIONS FOR REVIEW.

What is the first work of the eighth grade?

Why is the linen patch more difficult than the one of gingham?

What is the size of the linen patch model?

What is the first work on this model?

Why is the hemstitching of the ends and the hemming of the sides of the model left until the patches are completed?

How are the patches placed?

After the patches are set, and the hemstitching and hemming done, how is the model finished?

How are the measures of the girl's waist taken?

In what way is the girl's waist different from the child's waist?

How are the darts placed ?
How is the side form placed ?
How is the back slant obtained ?
How is the sleeve drafted and cut ?
How is the skirt cut ?
What is darning ?
How is a diagonal tear mended ?
Like what sort of weaving is the over and under darn ?
Like what weaving is the linen darn ?
Is the knitted darn like any kind of weaving ?
Is all patching done in the same way ?
How does the drafting of pants differ from the drafting of drawers ?
How does the blouse waist differ from a child's waist ?
What geometrical figure is used in cutting the different garments of this system ?

SPINNING AND WEAVING.

Spinning is the art of twisting together a number of filaments or fibers in such a manner that a thread or line of greater length than the single fibers of which it is composed is produced. So ancient is this art that nothing is known of its beginning. Herodotus, Ovid, and other classic historians tell of spindle and distaff spinning. The flax was wound about the distaff with one end inserted in a slit at the top of the spindle, which is a stick ten or twelve inches long. The weight of the spindle continually carried down the thread as it was formed.

A great improvement on the spindle and distaff was the hand spinning wheel. When or by whom this was invented is not known. An excellent thread was made with this wheel; but the process was slow and laborious, and as a consequence the weaving industry was very much circumscribed. The invention of the spinning jenny by James Hargreaves in 1764 revolutionized weaving as well as spinning. By substituting the mechanical for the manual process, one person could spin as much as twenty persons could with the spinning wheel. But the thread made by the mechanical process, while suitable for weft, was only fairly good for warp. It remained for Richard Arkwright to invent a machine, five years later, with which a thread suitable for all purposes could be made. But this was not the end. Samuel Compton, uniting the best points of the Hargreave and Arkwright machines, fixed the

Spinning.

creels of rovings in the frame, and transferring his spindles to a moving carriage, produced the spinning mule. Thus, from the crude beginning of spindle and distaff, has developed the time-old art of spinning, which now is accomplished with wonderful speed and very little manual labor.

"Weaving is an art," says Dr. Johnson, in his dictionary, "by which threads of any substance are crossed and interlaced so as to be arranged into a permanently expanded form." In all weaving, there are two kinds of threads used, one called the warp, and the other the weft. The warp, which is generally, but not always, the parallel threads, is mounted on the loom before the weaving begins. The weft is the thread that crosses and intersects the warp.

The first looms were two transverse bars attached to pegs driven into the ground. Between these bars the warp was extended. The weaver, sitting flat on the ground, put the weft under and over the warp with his hands, using no implement whatever. Then came the vertical loom, at which two weavers could work, although they used their hands only. Still better was the Grecian vertical loom. With this was used a rod which was both shuttle and batten, and which had a hook on the end by means of which the weft was drawn through the warp.

The development of this universally necessary art was very slow. Even as late as a hundred years ago crude looms were to be found in almost every farmhouse, and a large proportion of the making of cloth was an individual matter, as all but the very rich spun the thread for, and wove, such fabrics as they used.

Spinning Wheel.

Inadequate as were these looms of a century ago, in comparison with those used in the great factories of the present day, they were elaborate labor-saving machines compared with the crude, simple looms which are still used in India in making such exquisite fabrics as India muslins and cashmere shawls. These looms are simply two bamboo

rollers, one for the warp and one for the weft, and a pair of gear. Under a convenient tree, the weaver digs a hole large enough to contain his legs and the lower part of the gear. He then stretches his warp by placing his bamboo rollers a certain distance apart, and fastening them with wooden pins. The rest of the gear he fastens to a branch over his head. In two loops underneath the gear he inserts his great toes, which he uses as treadles. The shuttle with which he puts the weft through the warp is a large netting needle, which he uses as a batten to push each thread closely up against the last one put through.

Until 1733 the shuttle containing the weft was put through the warp by the weaver's hand. In that year, John Kay invented the flying shuttle, which is a

Grecian Vertical Loom.

mechanical device that takes the weft thread swiftly and evenly through the warp without as much as the touch of a hand. This machine enabled one person to do as much as two could accomplish by the old method.

Plain cloth is made by simply putting the weft thread under and over the warp. For fine cloth, the warp threads, which are very delicate, are placed so that they lie closely together, and the weft threads, which are equally delicate, are put in so that they lie as close together as the warp. The process by which the weft threads are made to lie close together is called "battening," or beating the weft up in place. Frequently, part of the warp and part of the weft are colored in such a way as to form checks, as in gingham, or simply stripes.

Indian Out-Door Loom.

Corded surfaces and an almost endless variety of effects are obtained by an arrangement which causes the weft to pass over and under two or three threads instead of a single thread of the warp. In making satin, which had its origin in China, the passing of the weft through the warp is so managed that a smooth surface is presented. What is known as three-leaf weaving is the simplest twill, and is where the weft passes over two and under one warp thread, giving the appearance of a succession of diagonal lines. Cashmeres, serges, and all kinds of goods with a twilled surface, are woven in this way, although the number of threads that are taken up or passed over varies in different kinds of cloth, as may be seen by raveling out a piece of twilled goods, and observing how the threads are placed.

To weave cloth in intricate and artistic patterns of various colors, a special loom is necessary. Such a loom was invented by Joseph Marie Jacquard, in Lyons, France, in 1801. It is really a combination of machines; and although simple, the results obtained are nothing short of marvelous. It was invented when Napoleon I. was Emperor of France, and hearing of it he sent for the inventor. When Jacquard arrived, the emperor said to him: —

"Are you the man who pretends to do that which God Almighty cannot do, tie a knot in a stretched string?" For answer, Jacquard produced his machine, and tied the stretched string. The emperor acknowledged that he could do what he had supposed was impossible, and awarded him a pension of a thousand crowns (twelve hundred dollars) a year.

Loop or pile weaving is where the weft is arranged in a series of loops, as in Brussels carpets. This kind of weaving is cut or uncut, as the case may be. Velvets of different kinds are woven in this way.

It was in 1790, at Pawtucket, R.I., that the first factory for weaving cotton cloth in the United States was established. Since then the most wonderful machines have been invented for weaving cloth rapidly and beautifully, and yet some of the finest work of this kind is still done by hand.

*9 7 8 3 3 3 7 4 1 8 3 5 9 *